cat and kitten
behaviour

AN OWNER'S GUIDE

Sarah Heath BVSc MRCVS

Sarah Heath qualified as a veterinary surgeon from Bristol University in 1988 and spent four years in mixed general practice before setting up a behavioural referral practice in 1992. She sees cases at monthly behavioural clinics at Bristol and Liverpool University Veterinary Schools and at private veterinary practices in northwest England. She is currently Secretary for the BSAVA affiliated Companion Animal Behaviour Therapy Study Group and Veterinary Officer for the Association of Pet Behaviour Counsellors. She is also Secretary of the European Society of Veterinary Clinical Ethology. Sarah lectures extensively on behavioural topics and contributes to the undergraduate veterinary curriculum at both Liverpool and Bristol Universities and the Royal Veterinary College. In addition to her clinical work, she writes regularly on behavioural topics for veterinary publications and popular magazines, has written a book called *Why Does My Cat...?*, and is involved in radio and television work on the topic of animal behaviour.

Dedication

This book is dedicated to my son Matthew – with all my love and thanks for everything you mean to me.

cat and kitten behaviour

AN OWNER'S GUIDE

SARAH HEATH BVSc, MRCVS

First published in 2001 by
HarperCollins*Publishers*
77–85 Fulham Palace Road
Hammersmith
London W6 8JB

The HarperCollins website address is
www.fireandwater.com
Collins is a registered trademark of
HarperCollins Publishers Limited

06 05 04 03 02 01
9 8 7 6 5 4 3 2 1

ISBN 0 00 710063 9

This book was created by
SP CREATIVE DESIGN
EDITOR: HEATHER THOMAS
ART DIRECTOR and PRODUCTION:
ROLANDO UGOLINI

PHOTOGRAPHY BY ROLANDO UGOLINI
Additional photographs by: Charlie Colmer,
David Dalton, Bruce Tanner, Cathy Gosling
(12, 58), Marc Henrie (29) and Marion
Rutherford (45, 47t).

Colour reproduction by Colourscan,
Singapore
Printed in Hong Kong

PUBLISHERS' ACKNOWLEDGMENTS

The publishers would like to thank the
following people and their cats for their kind
assistance in producing this book: Mark and
Sam Aldous, Karen Cleyer, Chris and Lesley
Fisk, Mel and Neil Gardner, Monica and Jim
McLaren, Sarah Palmer, Laura Riches,
Marion Rutherford and Jenny Thornton.

AUTHOR'S ACKNOWLEDGMENTS

I would like to thank my colleagues in the
behaviour world for their support and
encouragement. It is impossible to thank
everyone by name but a special mention goes
to Tiny De Keuster, Rachel Casey, Petra
Mertens, David Appleby, Jon Bowen and
John Bradshaw, who have been a constant
source of help, advice and humour over the
years. I would also like to thank Gwen Bailey
for introducing me to HarperCollins and
supporting the idea of a book on cat and
kitten behaviour and Cathy Gosling for
believing in the project. Writing this book
would not have been possible without the
assistance of Joanna McEwen who has kept
the practice running smoothly and fielded the
phone calls with true professionalism. My
family also deserve a mention for their love
and patience and I would like to thank
Matthew, David and my parents for being
there. Finally thanks to all the cats in my
life so far and to Gremlin in particular, for
helping to develop my love and appreciation
of this most fascinating species.

Note: Throughout this book, 'he' rather than
'he/she' or 'it' has been used. This is to make
the text easier to read and does not reflect on
the relative worth of male or female animals.
Cats of both sexes make good pets, and both
sexes have their own disadvantages and
positive attributes.

contents

Introduction

The rising popularity of the cat over recent years has resulted in it now being recognised as the number one pet in the United Kingdom. Its independence and cleanliness are amongst the principal reasons for its increasing profile as a companion animal. However, with the increasing feline population there has been a corresponding rise in the number of behaviour problems that are reported by their owners.

These behaviour problems can be very varied and medical conditions must always be considered when we are investigating cats that are behaving in an unacceptable manner. However, many of the aspects of feline behaviour that cause distress among the cat-owning public are related to the cat's natural behaviour. An understanding of feline society and the different ways in which cats communicate with one another will help owners to see things from a feline perspective and to put the behaviours in a feline context.

The cat originally evolved in the African bush, where territorial behaviour was vital for survival and individuals rarely met members of their own species unless they were very closely related. As a result they had to find ways of communicating that did not involve meeting face-to-face and this led to the use of marking behaviours that could be left behind for other cats to read at some time in the future. These marking behaviours are extremely useful for cats but when they occur

LEFT: *Cats are becoming more popular and are being seen as true companion animals.*

in locations that are unacceptable to the owners misunderstanding can quickly lead to a deterioration in the pet-owner relationship.

Cats have proved themselves to be one of the most adaptable species in the world and the story of their relationship with man is one of unparalleled contrasts of worship and hostility. Understanding the process of feline domestication puts a new light on the role of the cat in our society today and fosters a deeper appreciation of this most fascinating of species.

The cat-owner relationship can be very rewarding, but living in close proximity with another species is not without its problems. When we place demands on our pets in order to make

ABOVE: *Scent is very important in establishing and maintaining relationships.*

them fit into our world we need to take responsibility for helping them to adjust. Understanding the importance of kitten development is essential. Kittens are at their most receptive to socialization and habituation between the ages of two and seven weeks, and providing imaginative opportunities for them to meet a wide variety of people and experience a range of environments and situations is not an option but an obligation. If we take this responsibility seriously the cat will adjust more readily to domestic life and the cat-owner relationship will be enhanced.

Chapter one

The history of man and the cat

Cats are increasing in numbers around the world. They offer an outlet for human nurturing qualities while retaining a degree of independence that makes them fascinating to observe. Cat watching provides us with an insight into the behaviour of a species that has evoked extreme feelings of devotion and hatred throughout the ages.

The dictionary definition of domestication is to 'tame or bring under control' and most owners will be able to think of cats who would vehemently object to the implication that they are dependent on man for their survival. The cat has been described as 'an exploiting captive' and a 'carnivore that enjoys the company of man'. Of course, man has manipulated the cat to some extent and has bred for tolerance, affection and tameness, but for most cats life still involves a degree of freedom. For the non-pedigree cats, at least, this freedom makes it impossible for humans to be totally in control of breeding programmes and therefore the species has managed to escape the extremes of manipulation that dogs have gone through in terms of appearance and behaviour. In the pedigree cat world, there is a larger degree of control but even then history has not seen the diversity in shape, size or function that pedigree dogs have displayed.

In order to understand the way in which the cat and man co-exist today, it is useful to look back in history and study the relationship between our two species. No other domestic animal has had such a varied relationship with man. It has commanded extreme respect as a goddess, and extreme revulsion as an agent of the devil.

In many quarters, cats still evoke strong emotions of devotion and hatred and, just as we have seen throughout history, the very same feline qualities that appeal to the cat-loving public make them almost abhorrent to other sectors of society.

Cats in ancient Egypt

The exact date on which feline domestication began is still a mystery but the association between man and cat is believed to have begun in Egypt in about 4000 B.C. Art has offered much of the evidence of the development of this relationship and early Egyptian tomb paintings dating from 2000 B.C. depict the cat in unmistakable association with man. The art can be divided into two main types: firstly, those pieces depicting the cat in a marshland setting hunting for birds; and, secondly, those showing cats in more obviously domestic settings. One of the most famous marshland paintings was found in a tomb at Thebes and shows Nebamun hunting fowl in the marshes accompanied by a retrieving cat. It is widely accepted that this image of a cat being used to retrieve must be symbolic, since such behaviour does not form part of a feline's natural repertoire, and in this specific painting it is thought that the image represents fertility in a painting that shows people enjoying a plentiful supply of game.

Art depicting cats in a more formal domestic setting has been retrieved from a number of tombs in the Valley of the Nobles at Thebes and the cat is usually depicted beneath the chair of the woman in the painting, which is believed to be another symbol of fertility. Paintings from the Tomb of May dating back to circa 1450 B.C. show the cat tethered to a chair leg, and the famous wall paintings from tombs at the village of Deir el Medina near Luxor dating from circa 1275 B.C. also portray the cat sitting beneath the chair of the woman. The cats that are depicted often show striped tabby markings, suggesting they were still closely related to the African Wild Cat but the increasingly domestic settings suggest that the relationship with man began around the start of the Eighteenth Dynasty.

The formal relationship between man and cat may have been linked with the development of settled agriculture. Grain stores attracted vermin and the small wild cats were probably attracted closer to human communities by the easy supply of food. Those cats that were prepared to come closest would stand the best chance of catching more vermin, and thus a selection process for bold and confident cats was established.

OPPOSITE: *The beautiful Egyptian Mau is believed to be domesticated from a spotted subspecies of the African Wild Cat.*

Cat worship and domestication

There does not appear to be any evidence of domestication of cats outside Egypt and this geographical restriction is related to the status of the cat within Egyptian culture. For more than 1300 years the Egyptians regarded the cat as a deity, males being sacred to the sun god Ra and females to Bastet, the feline fertility goddess. Large numbers of cats were kept confined to temples to prevent them interbreeding with the local wildcat population. This may have been a factor in accelerating the process of selection for individuals who tolerated the proximity of man.

Despite the divine status of the cat and the devotion it commanded, strange rituals are believed to have been carried out, including mass sacrificial strangling of kittens between one and four months followed by the internment of their remains in a huge feline cemetery at the Temple of Bubastis on the Nile Delta. This temple was the focus of cat worship and the annual festival attracted hundreds of thousands of pilgrims paying homage to Bastet. In her original depictions she had the head of a lioness but feline domestication probably marked the time when her image was modified to that of the domestic cat.

The Bastet statue carried a four-stringed musical instrument – a sistrum – in one hand which was believed to sound out the rhythm of lovemaking. In the other hand was an aegis, or shield, depicting chastity, and a basket, which was believed to reflect the appearance of Bastet devotees. As well as the four strings of the sistrum, the statue incorporated four kittens sitting at its feet. Both of these references to the figure four represent high-level fertility, since female cats usually give birth to four kittens.

Although the cat retained its deity status things began to change, and by 1600 B.C. cats started to live outside of temples and were kept by the Egyptians as domestic pets. They still

ABOVE: *Cats appear frequently in the art of Ancient Egypt, most notably in statues of Bastet, the feline fertility goddess.*

commanded an unparalleled level of devotion; not only was it illegal to kill a cat but when a cat died all the members of the household would shave off their eyebrows as a mark of respect and enter a period of official mourning.

In 950 B.C. when Bastet became the national deity in Ancient Egypt, the cat was at the very peak of its period of devotion. Evidence of the important role of the cat throughout all sections of Egyptian society is provided by the range of feline mummies that have been retrieved. Elaborate linen wrappings were used to mummify the cats of the rich and famous whereas simple cloth mummies were created for the cats of the lower classes.

The worldwide spread of cats

Cats in ancient Rome

While the cat was fulfilling a role of divine rodent controller in Ancient Egypt the Romans were using ferrets and polecats for the same purpose, and it was not until A.D. 400 that the Romans began to adopt the cat to control rodents. Before that time, foreigners were considered unfit to keep cats and if any were known to have left Egypt they were rapidly captured and returned to their 'rightful home'.

The exact route for the spread of the cat is not known but it is thought that the movement of the Egyptian army in 595 B.C. when they were attacking Pelusia may have been a contributory factor in the transfer of cats across into Persia.

Freer movement of cats out along the trade routes is believed to have followed the banning of worship of Bastet by the Romans in A.D. 390, and the rapid spread of the cat throughout Europe is thought to have gone hand in hand with the expansion of the Roman Empire and the establishment of Roman settlements. Skeletons of cats have been found in some Roman villas that have been excavated in Britain, and the remains of fourteen cats were found at the Roman settlement of Tac in Hungary.

Cats in Asia

Before A.D. 1000 the cat had spread along the commercial routes between Asia and Europe and was established in China as a symbol of peace, fortune and serenity. The cat was considered to be a valued rodent controller and even to this day some

Asiatic communities attribute divine powers to the cat and believe that when a cat dies its soul speaks to Buddha on behalf of its owner who remains on earth. The legend is that the older a cat is and the less hair it possesses, the more good fortune it will bring but all is not rosy for the cat in the Buddhist religion and it does not appear on the list of protected animals in the original canons of Buddhism. This omission is rumoured to result from the cat falling asleep while on duty at the funeral of Buddha. Movement from China to Japan led to the cat becoming established in the imperial palace of Kyoto, and cats probably lived there from A.D. 999.

The arrival of cats in India is believed to have happened at a similar time to China, and the feline goddess Sasht, who was a symbol of maternity, has been compared to the Egyptian deity Bastet (see page 12). In the Hindu religion, there is an obligation for people to offer food and even hospitality to a cat.

Cats in mediaeval Europe

Once the cat had left the confines of Egypt its relationship with man was to change beyond all recognition and the roller coaster of a love affair between man and cat was to truly begin. When the cat first appeared in Europe it still had divine status and

Global cats

Having broken free from Egypt, the spread of the cat was rapid and in modern times their distribution has become global. Cats are believed to be most popular in the Antipodes, North America and Western Europe, but there are some cat-loving countries outside these areas, most notably Algeria, Israel and Indonesia. The extent to which cats are 'owned' varies considerably between countries, and in many Mediterranean cities there are large populations of strays, as many tourists can testify.

the sistrum that had been held in the hand of Bastet still accompanied it. Indeed, the sistrum is believed to have been the basis for mediaeval portrayals of the cat playing the fiddle. The ancient nursery rhyme 'Hey diddle diddle' refers to this association with the sistrum and also brings in elements from the cow goddess Hathor and the moon, which was believed to be connected with cats through the reported effect of the lunar cycle on their pupil size, and with women through the links between the moon and their monthly cycle. The side arms of the sistrum were believed to symbolize the horns of the cow goddess Hathor, and the roundness of the sistrum the lunar sphere. Hence the cat, the fiddle, the cow and the moon are brought together in those familiar words 'Hey diddle diddle, the cat and the fiddle, the cow jumped over the moon'.

Good fortune

As Christianity was established as the Roman Empire's accepted religion, the cat's divine status was lost but throughout the mediaeval period the cat was still believed to retain certain magical properties and many stories which have survived from this period refer to cats who brought their owners good fortune.

The well-known story of Dick Whittington was written in the seventeenth century but is believed to have its roots in the traditional stories of the mediaeval period when cats brought good luck to their masters. Certainly Dick Whittington did exist and he did make a fortune and become Lord Mayor of London but there is no evidence of him ever owning a cat. It is thought that the first picture of him holding a cat originated from the artist's belief that sudden good fortune should be linked with the magical intervention of a feline rather than the actual existence of such a cat!

Not all stories of feline magic gave such pleasant endings as that of Dick Whittington, however, and the human interpretation of feline magical powers was to lead to some thoroughly unpleasant practices, not least of which was that of interring cats in the walls of houses to ward off evil spirits.

Persecution of the cat

During the thirteenth century, the fortune of the cat took a dramatic dive, and as man became increasingly intolerant of his fellow man the cat also began to suffer from the inhumane manifestations of human ignorance. The magical powers of the cat were now believed to have a more sinister aspect and were widely thought to indicate an association with demons and with witchcraft. As a wave of hysterical persecutions against witches swept across Britain there followed a campaign against their feline companions.

Not only were people persecuted for what were believed to be their 'improper' relationships with cats but the cats also became victims, and in a period of unparalleled cruelty they were burnt to death in their thousands. Cats were roasted in cages above fires and the prolonged nature of their death was justified in terms of making the devil suffer. These practices continued until the mid-seventeenth century and the vicious abuse of cats was widespread long after the burnings stopped.

Demonic powers

It is ironic to discover that the very aspects of the cat that had led to its divine status in Egypt were now condemning it as a symbol of evil. The cat's reflective eyes, which are the result of a special feature of their retina called the tapetum, were believed by the ancient Egyptians to hold the rays of the sun and to symbolize deity status yet the same feature was believed by people in the Middle Ages to symbolize demonic powers. Other misinterpretations of feline biology added fuel to the fire and the cat's promiscuous lifestyle together with the scream of the queen as the male cat withdraws

BELOW: *The legacy of mankind's turbulent relationship with the cat lives on and even today black cats are linked with superstition.*

after mating were taken as clear indications that the animal was possessed by the devil.

During the reign of Mary Tudor in England, cats were burned as embodiments of Protestantism but ironically the same practice during the reign of Elizabeth I was justified on the grounds that cats were symbols of the Catholic church. In France, crowds would assemble in town squares to celebrate the festival of St John on 24 June, and captured cats would be thrown onto fires amidst the hysterical cries of the people who believed that they were liberating themselves from the devil. Churchmen and royalty condoned these practices and by A.D. 1400 the domestic cat was virtually extinct. However, there was a dramatic price to pay for this and with insufficient cats to kill the rats that carried the bubonic plague two-thirds of the human population of Europe died.

A change in attitude

Although religious persecution of the cat began to tail off towards the mid-eighteenth century, persecution and torturing did continue in the course of baiting sports and it was Louis XIV who prohibited cat-burning ceremonies and began the change in attitude to felines. However, it was not until the French Revolution that previous beliefs about cats were labelled as pure superstition and the practices of the preceding years seen to be cruel. Even then misunderstanding of the cat's natural behaviours continued and for a long time cats were more commonly found in feral groups scavenging from human garbage than living as house pets.

The cat's independence was seen as a sign of unreliability, and most people felt that an element of untrustworthiness was associated with the feline species. This situation continued for many years with the cat effectively living alongside man rather than with him.

A new era

In the mid 1800s Louis Pasteur announced his scientific discovery about disease transmission and people began to decrease their belief in illness as a punishment of God or a work of the Devil. At this time the cat's fastidious grooming habits earned it recognition as a shining example of hygiene, and the cat entered a new era. The keeping of pets was becoming fashionable and in 1871 Harrison Weir set out to change public attitudes by organising the first cat show in the UK. Through that event he sought to give the cat an air of respectability, which would allow it to take its rightful place, as a much-loved human companion.

Choosing the right kitten for you

When owners are considering the right sort of kitten to take on there are a number of very important questions that need to be answered. Although the different cat breeds and types do not vary in appearance to the same extent as the dog, it is still true to say that not every type of cat is suitable for every owner. Considering your own lifestyle and the way in which you want your cat to integrate into it will help you to make the right choice. Variations in personality, in activity levels and in certain aspects of behaviour can be predicted with certain breeds but in general the first things for a prospective owner to consider are the conformation, coat type and coat colour of their new pet.

What sort of appearance are you looking for?

Unlike the dog, which ranges in size from the tiny Chihuahua to giant breeds such as the Great Dane, the cat has a more uniform size and its shape does not vary to any great extent. In general terms, cats can be divided into three basic shapes:
◆ The slim, elegant oriental type
◆ The stocky, muscular British and American shorthairs
◆ The squat, flat-faced Persian or Exotic.

Of course, there are also some distinct structural differences in the cat world, produced by selective breeding, and the tailless Manx cat is a good example. Other breeds that carry distinct appearances include the American Curl, with its outward curling ears, the Scottish Fold, with its tightly folded ear flaps, and the Japanese Bobtail, with its curly-kinked tail.

Have you considered the issue of grooming?

Coat length is a very important consideration in the selection of your kitten since taking on a long-haired cat will involve you in a lifelong commitment to regular grooming.

Long-haired breeds

Many owners want a cat because of its ability to care for itself but the long-haired breeds have been developed to the point where they are dependent on their human carers to help them in the grooming process. For these cats, an introduction to daily grooming from a very early age is essential and if you have a hectic lifestyle with little time for routine chores I would suggest that long-haired cats are not for you. If you decide that you really do have the time and the commitment to take on a long-haired cat you will need to start the daily routine of grooming from the very first day. In fact, the kitten should have been introduced to the brush and comb before it reached seven weeks of age but this will have been the responsibility of the breeder. Once it arrives in your home, the kitten will need short regular grooming sessions and you will need to ensure that the experience is as pleasant and rewarding for your kitten as possible (see Chapters 3 and 5).

Semi-long-haired breeds

The semi-long-haired breeds of cat, such as the Birman and the Maine Coon, have been suggested as being something of a compromise but these breeds will still need considerable attention in terms of their coat, and grooming must be a regular event for these individuals as well.

Short-haired breeds

The short-haired cats remove the need for daily grooming by the owner and the more unusual Rex breeds and the Sphynx obviously do not require you to keep a brush and comb at the ready. However, grooming will still form part of the care routine for most cats and you will still need to look at your cat's coat on a regular basis. Just because a cat does not have a long coat it should not be looked upon as an easy option.

BELOW: *Daily grooming of long-haired cats is essential.*

Which coat colour?

Cats come in the most fantastic range of colours and this is certainly one of the criteria that new owners use when selecting their kitten. It has often been said that the colour of the coat gives some clue as to the personality of the cat but this has never been proven. Experience tells us that dark tortoiseshell cats are often fiery in temperament whereas ginger-coated felines are generally regarded as being laid back and friendly.

ABOVE: *The domestic shorthair is very popular and owners are afforded a remarkable choice in coat colours.*

Do you want a pedigree cat?

Over eighty per cent of the pet cat population in the UK consists of non-pedigree animals and most people think of the cat in terms of the domestic short-hair. However, there are a number of pedigree breeds available. In the dog world, ownership of pedigree pets is commonplace so why is there such a difference in attitude amongst cat owners? Selection of a pedigree animal usually allows some level of prediction over what the adult pet will look like and this is true for cats as well as dogs, but in the canine case behavioural differences are also likely to be marked due to the influence of function on dog breed selection; this is not as relevant to the cat. When a dog owner is selecting a breed they may have strict criteria in terms of function; they may even have a job for the dog to do. Cat owners are more likely to select in terms of appearance, and although personality may have some bearing on their choice a lot of potential owners will simply want a cat!

Another reason why there are less pedigree cats than pedigree dogs is the fact that kittens have traditionally been thought of as a free pet and some people find it incomprehensible that someone would pay for a kitten when so many are looking for good homes. However, times are changing and as the cat becomes the number one companion animal in many countries, its image is going upmarket and pedigree cats are on the increase.

Which breed do you prefer?

There are a wide variety of cat breeds for you to choose from and countless books are available to help you in the search for the breed that is right for you. In general the variations between the breeds are recorded in terms of conformation, coat type and personality but it is generally agreed that differences between individual cats are greater than the general variation between breeds, and descriptions of temperament and personalities within breeds are recognised as being broad generalizations. Here are a selection of some of the common, and not so common, breeds together with some information about their appearance and temperament.

Short-haired cats

British Shorthair
These are large cats and their broad chest, short sturdy legs and round head give them an unmistakable appearance. They have large penetrating eyes which are wide set and their ears are small and neat.

LEFT: The British Shorthair has a friendly, affectionate nature.

The tail is thick with a rounded tip and the coat is dense and plush in appearance. They have large rounded paws and can be found in a wide range of colours and patterns. The colours include blue, chocolate, cream, lilac, black, red and white. Patterns include the self or solid, the bi-colour, the spotted, the tabby and colourpoint. The American Shorthair has a larger head and longer legs and tail but it is seen in a similar range of colours and patterns. Both of these short-haired breeds are known for their easy-going nature.

Siamese
In total contrast to the British Shorthair, this breed is renowned for its sleek body and characteristic elegance. Its striking blue eyes endear it to many, and its fine legs, long tail and long straight nose give it an air of royalty. Its coat is short, sleek and

OPPOSITE: The Siamese's regal appearance makes it a firm favourite with many cat lovers.

silky and comes in a range of point colours from the most common seal to the less well-recognised lilac, cinnamon, caramel and fawn. The Siamese is the most vocal of the cat breeds. It craves human company and is described as active, intelligent, inquisitive and demanding.

Burmese

This is a popular breed, sharing many of the characteristics of the Siamese, including its demanding and human-orientated nature. It has a more rounded physical appearance with wide-set ears and a distinct nose. It has huge, appealing eyes and high cheekbones. Its coat is dense and glossy and as well as the original seal colour, Burmese are available in lilac, blue, platinum and champagne. There are also four tortoiseshell varieties: the brown, blue, chocolate and lilac torties. Although Burmese cats love to play and have been described as

extraverts, their interaction with people tends to be less vocal than the Siamese. They are athletic, intelligent and confident cats who have plenty of energy and want to be in the midst of whatever is going on.

Russian Blue

Not as numerically popular as the Siamese or Burmese, this cat is famous for its gentle and affectionate nature. It is medium-sized with a short-wedged muzzle, tall upright ears and prominent whisker pads. It has a unique thick, plush and upstanding coat (a 'double coat'), which makes the characteristic difference between this cat and the coats of breeds with a similar colour. As its name suggests, this breed only comes in one colour in the UK, although white and black are reported to be bred in other countries. When asked to describe the Russian Blue, a large proportion of American

BELOW: *With its curious nature and zest for life, the Burmese makes an ideal companion.*

cat show judges commented on the cat's pensive personality and also on its limited use of its voice. 'Quiet' and 'silent' are adjectives that are often used to describe this breed.

Korat

In common with the Russian Blue, this breed is only available with a blue coat but it is tipped with silver giving it a very characteristic appearance. The coat is short and even toned and there should be a noticeable sheen. This is a medium-sized cat with a muscular body and a characteristic heart-shaped face with large, luminous green eyes. Interestingly, the eyes are yellow to amber-green in both kittens and adolescents. The breed originated in Thailand and breeders, who are very keen to preserve its natural form, disallow any outcrosses to other breeds. It is described as a playful and intelligent cat and it is generally regarded as being friendly and quiet.

Bengal

One short-haired breed that has captured people's imagination in recent years is the Bengal. This is an average-sized breed which is said to bear a striking resemblance to the Asian leopard cat. It has a muscular body with strong legs and unusually large feet. The Bengal carries its tail characteristically low and its head is rather large in proportion to its body. The coat is stunning with a texture more akin to the pelt of the Asian leopard cat than the fur of a domestic pet and a rosetted spot pattern with rosettes formed by partial circles of spots around a coloured centre. The coat colour can vary with three accepted patterns.

The Bengal is renowned for being non-vocal although some individuals do have a peculiar gravelly voice and their personality is best described as quick, agile, intelligent and somewhat tenacious.

Abysinnian

Affectionately known as the 'Aby', this breed is characterized by its medium-length, short-hair topcoat, which lies close to the body and consists of hairs which are double or even treble ticked. This banding of the hairs with alternating dark and light gives the coat its unmistakable colour, and as a result of the resemblance to the coat of the wild rabbit this breed is sometimes referred to as the rabbit cat. The original colour of the Abyssinian is a ruddy brown ticked with black but over the years other colours have developed, including sorrel, blue, lilac and chocolate. The Abyssinian is often described in the literature as being highly strung and cautious but once within a stable home it is renowned for its need to maintain contact with the humans within the household. A lively and intelligent cat, the Abyssinian is reported by some to become very restless when confined and to be totally unsuited to indoor life.

Long-haired cats

Persian

Technically the only long-haired breed, the Persian is also known as the Longhair and there are different types within the breed. The differences between them are subtle and relate to the colouring of the coat, the coat type and also the shape of the head and its features.

The Pekeface Persian is only recognised in the show world in the United States and it is interesting to note that there is considerable variation between the conformation standard for the Himalayan or Colourpoint in the UK and in the USA. The most striking thing about the Persian group of cats is their coat which should be long and thick with a soft undercoat. The body of the Persian type is cobby with short thick legs and a bushy tail. The face is another characteristic of the breed and it is true to say that this has altered significantly since the Persian first appeared at Harrison Weir's cat show in 1871.

Today's Persian has a much flatter face with smaller ears and large, round brilliantly coloured eyes. The solid coat colours of the Himalayan range from black, blue and chocolate to lilac, red and cream and there is also a range of torties. The colourpoints show colouring at the ears, stockings and tail which are darker than the whole body colour, and it has been suggested that the influence of Siamese blood makes these cats more lively and inquisitive than their solid colour counterparts. The Persian is considered to be laid back and easy-going and the combination of its affectionate personality and long coat has made it very popular as an indoor breed.

LEFT: *Persians are popular domestic pets but their luxurious coat needs a great deal of attention and this is not a breed for everyone.*

Semi-long-haired cats

Birman

This is a medium-sized breed with a solid body, rounded head and characteristic blue eyes. The face has a so-called Siamese mask and the shape of the head is considered to be of prime importance. The skull should be broad and rounded with the forehead sloping back so that there is a flat spot in front of the ears. The coat is semi-long and is silky throughout, lacking the woolly undercoat seen in the Persian. The four white paws are the hallmark of the Birman and give it a very distinguished and unmistakable appearance with the front legs being gloved in white and the hind ones being white at the front and with white gauntlets extending to the hock at the rear. The point colours include seal, blue, lilac, red, chocolate, tortie and tabby. The Birman is reported to be a sweet-natured breed which is very affectionate towards owners and judges alike and it is generally regarded as an ideal family pet.

Maine Coon

This is described as the largest feline breed but the size can vary enormously and the very large cats are definitely extremes of this breed. In general, the conformation of the Maine Coon includes a muscular, broad-chested body, a large wedge-shaped head with a medium-length, slightly concave nose, and a long flowing tail. The coat is dense and shaggy but should be longer on the stomach and trousers and shorter on the shoulders giving a characteristic ruff and bib of long hair. The coat requires a lot of attention but it is ideal for protecting the cat from harsh winter weather and the Maine Coon is very much an outdoor cat. During the summer, in countries where temperatures do really rise, the Maine Coon can lose most of its ruff and trousers. The eyes of this breed are large and expressive and the temperament is described as friendly, playful and companionable. They have a characteristic chirping voice which endears them to their owners and are often noted for their habit of sleeping in unusual places.

Norwegian Forest Cat

Another large solidly-built cat, the Norwegian Forest cat is renowned for its hard textured, water-repellent topcoat which covers a dense more woolly undercoat and offers the cat maximum protection in wet and cold weather. The coat has developed in response to the cold climate of this breed's native Norway and it is generally accepted that this is not a breed to be kept indoors. Not only is the coat of the Norwegian Forest cat developed for an outdoor life but the breed is also known for its hunting prowess and its climbing ability, both of which make it an adventurous and independent breed. However, there is also a friendly side to the personality and its adaptable nature makes it a rewarding companion.

Somali

The Somali is closely related to the Abyssinian differing only in its coat type, which in the Somali is a dense double coat with very fine densely packed hairs which are ticked and banded alternately light and dark. It is neither cobby nor oriental in shape and its face shape rests somewhere between the rounded British Shorthair and the pointed Siamese. The coat on the head is short and over most of the body is classed as medium length but the Somali is well known for its ruff, breeches and brush (the coat on the tail). The tail is a characteristic of the breed and is long and fairly thick at the base. This is a very extravert breed with a tendency to be vocal but not noisy.

Turkish Van

The Van's behaviour has made it remarkable in the cat world, and its love of water is now legendary. Once known as the Turkish Swimming cat,

BELOW: *The friendly Norwegian Forest Cat loves to be outdoors hunting and climbing.*

it originates from the Lake Van area of southeastern Turkey where it was reputed to enjoy bathing in the warm shallow waters. It is reported to take to the bath when there is no lake available for swimming! A sturdy cat, it has a long muscular body and strong legs. Its long silky coat is cream in colour over most of the body with an auburn tail and facial markings. It has a short muzzle and straight nose, large feathered ears and amber or blue eyes. It is renowned for its melodious voice and home-loving personality.

Designer cats

Although selective breeding has not played such a large role in the development of cat breeds as it has in dogs there has been a recent trend for the cat to be manipulated by man for more extreme appearances. This has been seen within some of the well-established breeds, including the

ABOVE: *The Turkish Van is famous for its love of the water and of swimming.*

Persian and Siamese, where selection has led to increasingly flattened faces in the former and thinner, bony conformations in the latter. In addition, there has been active selection in some quarters which has resulted in the arrival of more extreme breeds with very distinct alterations in both appearance and behaviour. In many cases, such extremes can have profound effects on a cat's lifestyle and its ability to communicate and can raise serious questions concerning feline welfare. A good example of altered communication can be seen in those breeds which have extreme ear conformations, such as the Scottish Fold and the American Curl.

Other examples of pedigree cat breeds where extremes of appearance have been actively encouraged are the Sphynx and the Munchkin.

Other things to consider

Will it be a queen or a tom?

The choice of sex is not considered so important in cats as it is in dogs and one of the reasons for this is that the majority of domestic cats are neutered before they reach puberty. This means that secondary sexual characteristics rarely need to be taken into account and the differences between the sexes remain minimal. Of course, those owners who are thinking of breeding with their cats or who have strong feelings about keeping their pet entire will need

BELOW: *Cats still remain very close to their wild type in their behaviour. The hunting instinct is strong in most kittens.*

to give the sex of the kitten some thought and also ensure that the cat is accurately sexed before they take it home. Many owners mistakenly believe their pets to be one sex and only find out their error when the pet needs veterinary attention. It is true that sexing very young kittens can be difficult and errors are often made. However, as the cat gets older the differences become more obvious and mistakes should decrease.

The issue of compatibility between the two sexes is one that may need to be considered when two kittens are being taken on at the same time or when a second cat is being introduced into an already established feline household. The general rule when acquiring two kittens is that it is preferable to take littermates of the same sex but, once again, pre-pubertal spaying makes this less of an issue. On the other hand, when taking in a new kitten it is usually easier to introduce a kitten of the opposite sex to the resident cat since it is less likely to be seen as a competitor even when the established cat is already neutered.

Do you like the sight of small prey in your house?

Hunting is a fact of feline life and all cats have an inner drive to seek

ABOVE: *Play is very important in kitten development, and social play between kittens can sometimes appear quite rough.*

and find suitable prey. However, some cats are undoubtedly more competent in this area than others and there is a belief that some of the hunting prowess is inherited while other aspects are learned. It therefore makes sense for those owners who hate the sight of small presents under the kitchen table to select kittens from poor hunting parents and to restrict their access to live prey during the early stages of development. Farm kittens are probably the least suitable pet for such people but the average moggie is also likely to have inherited a degree of hunting prowess and selecting a pedigree cat would probably be a better option in these circumstances.

One cat or two?

There is an old adage that two is better than one and when owners are taking on a new kitten there is a lot to be said for this approach. For many years cats were thought of as solitary creatures but a lot is now known about feline social structure and whilst cats certainly hunt alone it has been well demonstrated that they benefit from feline company. The advantages of two kittens are mainly related to the opportunity for social interaction, including play and

LEFT: *While some cats enjoy close physical contact with their owners, others prefer to retain their independence.*

lead to conflict and this is more of a consideration when obtaining kittens from different litters.

The best approach for taking on more than one cat is to obtain littermates since cats naturally live with those individuals that are related to them. Littermates will already be familiar with each other and that familiarity can be beneficial in reducing stress during the first few days in the new home. Of course, taking on two kittens should not be seen as an easy option and you will still need to spend time with each kitten to socialize it successfully and provide it with adequate care. Every long-haired cat will need the same high level of care in terms of grooming, and taking on two kittens will simply double this workload.

Do you work full time?

Cats have increased in popularity dramatically over the past ten years and one of the reasons for this is that they are believed to fit more readily into the busy lifestyle of modern man. Working full time is not such an issue for the cat owner as it is for those who share their lives with a canine companion, and it is generally accepted that the independent feline

mutual grooming, but many owners find that having more than one cat in the household gives them a unique insight into feline communication and behaviour which is priceless. Kittens can benefit from the companionship of another member of their species in many ways but it is important to consider the personality of each cat. In some cases, differing personality types can complement one another, and a shy and retiring individual may be encouraged to interact with a more boisterous and outgoing personality. However, differences in personality can also can see to its own needs and benefit from a relatively solitary existence.

Where owners still feel guilty about leaving their pet for hours at a time this guilt is often eased by taking on two cats but, in some cases, the close bond that the cats form with one another leads to problems of dissatisfaction on the part of the owner, who sees their opportunity for love and affection from their pets diminished. Certainly it has been shown that cats will preferentially spend time in the company of a member of their own species when one is available, but this does not preclude the owner from enjoying a fulfilling relationship

BELOW: *Better veterinary care has led to more feline geriatrics with a laid-back attitude to life.*

with both pets, and in most cases the cats are able to gain benefits from the company of both cat and human.

Cats certainly adjust well to life with a lower level of owner input than their canine counterparts, but it should be remembered that many of the pedigree cat breeds are renowned for their love of human company and these cats show a real desire for human interaction. As a consequence, some of the breeds are not suited to life with a working owner, and while separation-related behaviour problems are not common in cats, they certainly are being reported in significant numbers in some breeds. Matching the lifestyle of the owner to the demands and expectations of the cat should help to reduce these problems.

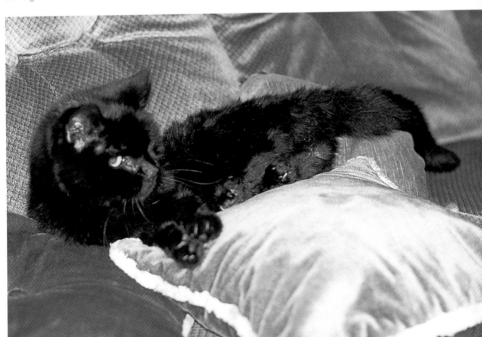

The right source for your pet

Once the decision has been made regarding the type, breed and number of kittens that you wish to own, you need to spend some considerable time selecting the right source for your new pet. If you are taking on a non-pedigree cat you will probably be able to find a kitten from a domestic background where the family pet has got out before it was neutered and taken its owners by surprise. Other common sources of non-pedigree kittens are farms, welfare societies and even pet shops and in some cases people take in unwanted kittens which have begun life in a feral situation.

In many cases, owners do not exactly select non-pedigree kittens but rather take them on in order to offer them a good home. No money changes hands but this does not mean that you should own any kitten without considering whether it is the right cat for you. The same considerations should apply when you are selecting

a non-pedigree kitten as when you are spending considerable sums of money on a pedigree one.

In general, the rule is that kittens need to come from healthy, hygienic and stimulating environments and need to have had the opportunity to socialize with people and to habituate to a domestic environment (see Chapter 3). Pet shops are not a suitable source for a kitten as they do not offer any background information on these animals and there will not be any opportunity to meet the mother, father or, in many cases, the littermates. Disease is a real

RIGHT: *Take care when choosing a kitten and don't be tempted to get one just because it looks cute.*

risk with kittens that have been taken from their litters and mixed with others, and early weaning and lack of maternal care can also lead to problems. Taking on sickly kittens just because you feel sorry for them may be commendable in some ways but owners often find that they pay a heavy price for their decision in terms of heartache and veterinary bills, and it is always best to let your head rule your heart in these situations.

Feral cats and strays

Feral cats are those that live a wild existence and only tolerate the proximity of humans in order to gain access to vital resources such as food. Without human interference, breeding is obviously uncontrolled in these cats and kittens are available in large numbers but their suitability as pets is questionable. It is important to make the distinction here between a stray and a feral cat as the former will have started life as a pet cat and become lost either accidentally or through being abandoned while the feral will never have lived in close contact with people.

Kittens from strays may well make good family pets since the parents will have been socialized and will often be very tractable and friendly individuals but feral kittens who have spent the first crucial weeks of life devoid of human contact are

Rescue centres

When kittens are brought into a rescue centre from unknown sources, their history is often vague and there may be problems involved in rescuing them. However, in many cases the pregnant queens are abandoned by their owners who don't want the responsibility of a litter of kittens. Such kittens can make ideal pets but you'll need to visit several centres and ask about the way in which the kittens have been handled before making a decision. Provided that you select a reputable rehoming organisation, the level of socialization and disease control will be good, and you will probably be the one being vetted to see if you are suitable as an owner.

likely to be nervous individuals prone to displays of defensive aggression. Feral kittens will have very strong instincts of self protection and they will often hide away from people which makes it even more difficult to get to them at an early age and try to socialize them. The sensitive period of development when kittens are most susceptible to the positive effects of handling is very short (see Chapter 3), and if feral mothers successfully hide their kittens from human gaze until they are past the seven-to-eight

ABOVE: *In the company of people, feral cats are prone to displays of defensive aggression. This behaviour can be traced back to their isolation from humans in the crucial early weeks of life.*

week stage of life socialization can be a very long, tedious and often unsuccessful exercise. Certainly some people with a great deal of skill and patience can work very effectively with feral kittens but in general this sort of rearing should be left to the experts and the average cat owner should look elsewhere for their new kitten.

Selecting the breeder

There is tremendous variation in the service provided by breeders and if you are taking on a pedigree kitten it is very important to take time to select a suitable one. The first thing is to locate some breeders of your chosen breed and this is best done by contacting the Governing Council of the Cat Fancy for a list of registered breeders. You may also find some information in breed magazines and general cat magazines, and you can visit a local cat show.

Once you have the details of a few breeders the process of selection can begin. If you want your kitten at a specific time you will need to check on availability of kittens. Cats have a breeding season and usually produce offspring in the spring and summer. The shortening of day length brings the queen out of season, since breeding in seasons of longer day length will ensure better hunting conditions to supply food for the young. Since non-pedigree queens usually go out and find their own mates, kittens are usually only available at certain times of year but this is not as true for the pedigree cat for whom breeding seasons may have been affected by years of selective breeding or may be manipulated by housing breeding queens in conditions of artificial light.

Visit the premises

Once you have established that kittens are available you will need to visit the breeder's premises. It is well worth taking time to look around the whole establishment as well as spending time with the litter in question. The cleanliness of the housing is important, and you will need to ask about the programme for prevention of infectious disease, but remember that complexity and stimulation are also essential for young kittens and therefore immaculate catteries may not be the best source for your new kitten if they are devoid of activity.

Meet the litter

If the premises are promising, the next step is to spend time with the litter and to ask to meet the queen and, if possible, the tom since they will both have an influence on the temperament of their offspring.

In most cases, the breeder will want to ask you a number of questions about your suitability as an owner of one of their kittens and such interest is usually a very good sign. You should remember to ask breeders about the socialization and habituation programme for their kittens, and if they are up to date with the latest ideas on kitten rearing they should also want to know what steps you will be taking to socialize and habituate your kitten in its new home.

What should you look for in a kitten?

When you are looking at individual kittens and trying to make a choice, there are a number of things that you need to bear in mind.

Appearance

Obviously appearance is going to be a big factor in your decision when choosing a kitten, but remember that this is not the most important one and in the case of a pedigree kitten you may find that a slight defect in appearance as far as the show world is concerned could give you the ideal opportunity to obtain a pedigree cat at a much reduced fee. In most cases, the breeder will insist that

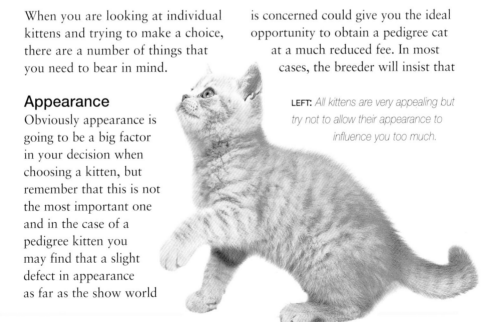

LEFT: *All kittens are very appealing but try not to allow their appearance to influence you too much.*

you sign an undertaking to have the cat neutered in order to prevent you from breeding and passing on the fault but for owners who are seeking a family pet this is not a problem.

Temperament

Temperament is very important when making your selection and your own lifestyle and family composition will have some influence on the best sort of cat for you. There are two recognised types of cats according to the behavioural research:

◆ Cats that have a high requirement for social contact
◆ Cats that are quite content to be on their own.

Obviously the latter is unlikely to be suitable as a pet for a young family where children want to play with their pet but it may be fine for a hard-working single person who is rarely at home and wants an independent cat. Those cats with the low requirement for company have been shown to spend even less time with people if attention is forced upon them and therefore these cats will settle better with busy people who are happy for the cat to take the initiative. Most people will have a preference for the high-requirement cat but even within this group there will be tremendous variation in the level of interaction that they demand and this will be determined to some extent by breed characteristics. The general rule for most pet owners is to select a kitten that reacts well to its littermates as well as to people and is confident and outgoing in new surroundings.

BELOW: Contented and relaxed kittens are likely to make well-adjusted adults, but a certain level of curiosity and mischief is also desirable.

Health

This is the other issue that you must consider when obtaining your kitten and while it may seem obvious to say that kittens that are clearly unwell should not be chosen, it is surprising how often sentiment takes over and people take on sickly kittens in an attempt to offer them a chance.

You can give the kittens a general overview, checking for weepy eyes, sticky ears and matted coats, but for most people the best way to ensure that the kitten is healthy is to have it checked over by a veterinary surgeon. This is usually done as soon as possible after you take the kitten home and breeders and rescue societies alike should be prepared to take kittens back if they are found to be unwell. In addition to checking the health status of the specific kitten, it is also sensible to enquire about the health of the rest of the litter, and when taking on a pedigree cat you will want to know about the disease status of all the cats on the premises. Infectious diseases are an important consideration in cats and reputable breeders will be happy to discuss the disease status of their stock with you.

Taking your kitten home

The general rule is that kittens should be obtained as soon as possible, but there is an important balancing act between leaving the kitten in the litter

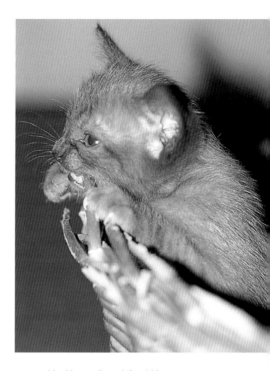

ABOVE: *Healthy, well-socialized kittens are very inquisitive and love to play with toys and other objects as well as their littermates.*

long enough for it to benefit from feline education and bringing it into its new home soon enough for it to benefit from its sensitive period of socialization. It is the latter part of this equation that is the more difficult for the kitten since the sensitive period is very early compared to the puppy. It is therefore important that the breeder starts the programme of socialization and habituation when the kitten is still with them.

Eight weeks of age is generally

ABOVE: *This thirteen-week-old Egyptian Mau kitten is ready to leave the breeder's premises but how it's been reared and handled will significantly affect its reaction to its new home.*

regarded to be a good compromise and most kittens obtained at this age adjust well to their new homes, but those owners acquiring pedigree kittens will probably find that they are not allowed to take delivery of their new pet until it is approximately twelve weeks of age. This is due to recommendations from the Governing Council of the Cat Fancy who suggest that pedigree kittens should be fully vaccinated before they leave the breeder's premises and should also be litter trained. There may be good reasons for this but, from your point of view, the important thing to remember is that your kitten will be well past the most sensitive period for adjusting to new environments before you take it home. It is therefore all the more important to ensure that the breeder's premises have provided complexity, novelty and challenge for your kitten thereby preparing it for life in a domestic setting.

Will your cat go outside?

When, after years of persecution, the tide of popularity finally turned for the cat and people began to look upon it as a suitable companion, the majority of cats were still kept at least in part for their hunting skills. The majority of these non-pedigree cats lived in rural areas where they provided first-class rodent control in return for food and shelter. Free access to outdoors was an unquestionable part of the package.

However, times have changed and cats are now expected to live in a variety of circumstances, some of which make free access to outdoors either impractical or undesirable. Owners who live in high-rise flats may be unable to arrange for easy access to outdoors, whereas those who live on very busy roads may feel that the risk of traffic accidents is too high and that their pet would be safer confined to barracks.

Whatever the reason for keeping a cat indoors it is absolutely essential that this preference in terms of lifestyle is taken into account when choosing your kitten. Not every cat is suited to an indoor life and while some pets will adjust well, provided that the owner takes the time to ensure that they make the necessary behavioural provisions (see Chapter 5), others will find confinement far too stressful for it to be fair. In general terms, the best kittens for an indoor life are those that have low-activity levels and are renowned for their affinity to people. Certain breeds offer these characteristics. Cats that have been born into feral situations and those whose parents are proficient hunters will not adapt well to indoor life. If a kitten has already tasted the outdoor life, bringing it into a totally indoor existence can lead to problems for both cat and owner.

BELOW: *Some indoor cats are content just to watch the world go by through a window while others suffer stress from being confined.*

Chapter three

Kitten development

M any people are familiar with Rudyard Kipling's words: 'Show me the child and I will show you the man'. There can be no doubt that early childhood experiences can shape and direct humans as they develop into adults. The same is true for the kitten and when trying to understand why an adult cat is responding in a certain way it is often helpful to look back at the early history of that animal and at the lifestyle and experiences that it received in its early weeks.

Before birth

Although development is often thought to begin at birth, other influences before then will also play some part in the behavioural development of kittens. Whilst human input into the pre-natal development of non-pedigree cats may be minimal it is important to take time to carefully select pedigree breeding cats with reference to their behaviour as well as their appearance.

Behavioural research has shown that the temperament of the tom cat has a marked effect on the behaviour of his kitten. Since most kittens never meet their father it can be assumed that the effect is genetic. Tom cats who are friendly and confident are likely to have kittens of a similar disposition, and it has been suggested that it is a so-called boldness trait that is being inherited. These bold kittens are not only confident around people but they have also been shown to be less stressed by confinement in a cattery or in a veterinary hospital and to be less nervous of unknown objects and situations they may encounter.

Of course, the father is not the only parent affecting the personality of the kittens and the mother's contribution is equally important, but it can be assumed that her influence will be transmitted to the kittens in a combination of genetics and example since she will have direct contact with her litter while they are developing.

The queen during pregnancy

In addition to the consideration of the genetic influences of the queen and tom it is also important to take into account the effect of the queen's health and mental condition during pregnancy on the subsequent behaviour of her kittens. It has been shown that experiences inside the uterus can have profound effects on the ability of young animals to cope with stress and this is due to an influence on a certain part of the brain during development. Kittens from queens who have been relaxed and contented during pregnancy are likely to be better able to withstand normal everyday stress when they are born, and paying attention to the mental wellbeing of pregnant queens as well as to their nutrition is very important. Obviously this is the role of the breeder and as a new owner you may not even have access to the relevant information in order to assess how the pregnancy has affected your kitten's development. However, you may be able to gain some insight by spending time at the breeder's premises and watching the behaviour of other cats who live there and it is always worth asking questions about the pregnancy to see if you can glean any information.

ABOVE: *It is important for the queen to be relaxed during pregnancy. This cat is enjoying a stroll in the garden a week before her due date.*

Kitten development

As far as a new owner is concerned the time when they will have the opportunity to influence the development of their kitten's behaviour will be once it has arrived in their new home but unfortunately much of the important development has already occurred by this time and therefore owners do need to ensure that their kitten has had the best start in life in behavioural terms before they take it on.

Kitten development is traditionally divided into four stages and these can be helpful in understanding when a kitten is most likely to benefit from particular activities or interactions. Obviously these stages are flexible and whilst age ranges are often stated it is important to remember that none of these are set in stone and every kitten is an individual. The stages give us an easy framework within which to work and guidance as to how we should interact with our kittens at various points in their development. However, when kittens miss out on the appropriate interactions all is not lost and mistakes can be rectified.

Stage 1: Neonatal period

The neonatal period runs from the moment of birth to approximately ten days to two weeks of age. The most noticeable activities of kittens at this early stage of life are sleeping and eating and the queen is the centre of her offspring's world. They are

ABOVE: *Very young kittens are dependent on their mother to stimulate urination and defecation.*

totally dependent on her for survival at this stage and are even unable to urinate and defecate without her assistance. She licks the underside of the kittens in order to stimulate the bowels and the bladder and this makes very good evolutionary

BELOW: *During the neonatal period the kittens are totally dependent on their mother for survival.*

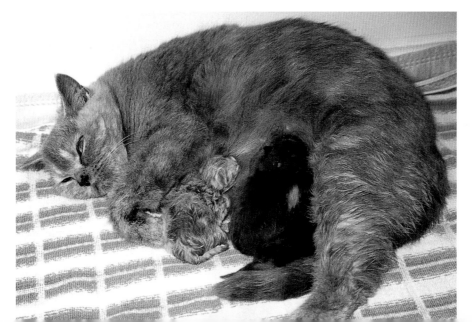

sense since it ensures that she is always present when her kittens go to the toilet and she can clean up the mess straight away. This ensures that urine and faeces do not threaten the hygiene of the nest area and, in the wild, will also decrease the risk of predators detecting the nest site.

Feeding

Although blind and virtually deaf at birth, kittens still manage to find their way to their mother's teats and they do so by using a combination of touch, smell and detection of warmth. The so-called rooting reflex drives kittens to push their head into warm places and thereby enables them to find food. Although unable to walk at this stage, kittens can drag themselves in the direction of warmth but this is a tiring activity

and they appear to be exhausted at feeding time when they have hauled themselves up onto an available teat. Once in contact with the teat a sucking reflex is triggered and kittens will also tread with their front paws on the area around the teat in order to stimulate milk flow. This so-called 'kneading' behaviour is often retained into adulthood and many owners will recognise this response when their pet is sitting on their lap. As the kittens develop, purring begins to be incorporated into the feeding routine and it is now thought that this has the combined effect of stimulating milk flow through its effect on the mother and also of communicating to the other members of the litter that food is available.

Body heat

Young kittens in the neonatal stage cannot control their own body temperature in the same way as adults and therefore cuddling up to the queen and the littermates is an important way of maintaining body heat. Separation from the litter is therefore potentially life threatening and even kittens as young as this will cry if they loose contact with their mother or littermates.

LEFT: *Kittens are able to locate their mother's teats using a combination of touch, smell and detection of warmth.*

ABOVE: *Cuddling up to the queen and its littermates is an important way for a kitten to maintain its body temperature.*

Reflex mechanism

During the neonatal period it is not unusual for the queen to move her kittens and when she does so she will carry the kittens by the scruff. This triggers a reflex response from the kittens who curl up their tail and back legs and go limp in their front legs. While they are in this position the kittens neither struggle nor cry and it

RIGHT: *Kittens should be handled as early as possible, but do so in a calm manner, taking care not to disturb and distress the queen.*

is thought that the reflex is a way of silencing the kittens and enabling the queen to avoid attracting any potential predators when she is moving the litter. The very same mechanism is triggered when an adult cat is scruffed during restraint and this can be extremely useful when you are dealing with a fractious individual or an animal which needs veterinary attention.

LEFT: *Picking cats up by the scruff can be a useful way of handling them but should never be used as punishment.*

Stage 2: Transitional period

Between ten days and three weeks of age the kitten is in its transitional period and at this time there are significant advances in its mobility and in the quality of its eyesight. The ears also open during this period and the kitten will start to investigate the solid food that is eaten by its mother. Probably the most significant development at this age is the onset of weaning, and although most people think of this in terms of an alteration in diet from liquid to solid it is also essential that we understand the behavioural significance of this process. The process takes place over a number of weeks and may even extend into the socialization and juvenile periods of development.

The importance of weaning

From the age of three weeks, the kittens find it increasingly difficult to gain milk from their mother and she will begin to get up and move away in the middle of a suckling session and start to ignore her kittens' demands for her to lie down and give them food. She will spend more and more time away from her kittens and when she does settle down with them she

LEFT: *At the start of the weaning process, kittens are on a continuous schedule of reward and the queen is content to oblige.*

will often lie down in such a way as to make the teats inaccessible. Obviously all of this behaviour is important in order to encourage the kittens to investigate alternative food sources but it is also essential in teaching them to deal with situations in which reward is not forthcoming.

This is the first time in their young lives that the kittens have experienced the emotion of frustration, and it is also the first time that they have found themselves in conflict with their mother. Up until now she has always responded to their demands with a generous supply of milk but her behaviour now is designed to withhold or withdraw that reward. This is a very significant turning point in the behavioural development

ABOVE: *Kittens are social animals and the presence of littermates provides security.*
BELOW: *As the weaning process progresses, the kittens' attention is diverted from their mother and onto independent food sources.*

of her young and it signals the start of a new independent life for the kittens in which their mother, the queen, is no longer all important.

ABOVE: *Scent is important in maintaining feline relationships at all stages of life.*

BELOW: *As kittens develop, their focus turns towards object-orientated play.*

Alterations in play

While the process of weaning turns kittens' attention onto solid food and onto the detection of prey there are significant alterations in the type of play that kittens engage in at this age. Until now most of the playful interactions have been with the queen or with littermates and this form of play is described as social. However, at the time of weaning the kittens will begin to show more interest in objects in their environment and they will start to act out the behavioural sequences associated with hunting. This so-called object play is believed to be important in the development of hunting skills. The presence of an experienced hunting mother is believed to increase the kitten's opportunity to observe the appropriate behaviours and thereby develop its own hunting prowess. Locomotory play is also important in developing balance and agility, which will be so important in adult life.

Stage 3: Socialization period

One of the most important phases of development for kittens in terms of their behaviour as pets is the so-called socialization period. During this time kittens learn very important lessons about their environment, both social and physical. In addition to learning how to relate to their own species, they also develop an

ABOVE: *Play provides an ideal opportunity for kittens to investigate the environment and to interact with their littermates.*

ABOVE: *Scratching is a very important feline behaviour, so you should ensure that the scratching facilities are fun to use.*

acceptance of others, and they form a view of the world which will strongly influence their behaviour later on as an adult. The socialization period is believed to last from around the second or third week of a kitten's life to the seventh week but, as with all developmental periods, the boundaries are flexible and there will be a considerable amount of individual variation.

The importance of feline company

Learning to be a cat involves a great deal of play behaviour with littermates and although the queen is obviously present at this time it appears that play between kittens is very important. Kitten play is rough and owners will comment on the fact that the attacks can seem quite severe but it is by inflicting injury on others that kittens learn to inhibit their aggression and to experience the consequences of their actions when they fail to do so. Although very young kittens will make physical contact with their opponents as they get older they learn to swerve out of the way at the last minute and limit the potential for injury by doing so. Play with the queen tends to be more predatory in nature, and it appears that she uses these play times to teach her offspring some important hunting lessons whereas the kittens use their playful interactions with one another to learn how to communicate.

Introducing other species

As well as learning to be a cat and to be able to communicate with fellow felines the kitten that is destined for life as a domestic cat also needs to learn about other species with whom it is likely to share its world. For most cats the two most important species are man and dog but there is a limitless potential at this stage of development and I have met cats that converse successfully with species as diverse as pigs, penguins and deer. Provided that kittens come into contact with other species while they are still in their most sensitive period for socialization they can learn to accept these species and to live with them in harmony. For many kittens, however, their lifestyle at this vital point in their development prohibits them from meeting other species and, as a consequence, the myth of the natural rivalry between cat and dog is thus perpetuated.

Learning to relate to people

For the majority of pet owners the most important consequence of this period of kitten development is the reaction of their adult cat to people and the importance of correct interaction both in terms of amount and style cannot be over estimated. Obviously the responsibility for getting this right rests primarily with the breeder or the people who are caring for kittens at this sensitive age and this is one of the most important considerations when selecting a source for a new kitten (see Chapter 2).

Research into the effects of handling kittens at this age on their subsequent behaviour toward people has shown that there are a number of factors to consider and the most important ones are the amount of handling, the frequency and style of handling and the number of handlers.

Kittens need diversity

As well as needing at least four different people to handle them at this sensitive age, kittens also require a diversity of types of people in order to assist in the generalization of human acceptance and those kittens that have been handled by young and old, male and female, noisy and quiet people will usually make the most bomb-proof adults.

It has been shown that kittens need to receive up to one hour of human handling daily during these important weeks of development and that the benefits are increased if it is delivered in short, frequent sessions. Kittens should be lifted up during the handling process and then gently restrained as this will prepare them for being picked up and cuddled by their owners later on in life.

Handling the kitten all over its body and examining its eyes, ears, mouth and coat on a daily basis will not only make veterinary care much easier when it reaches adulthood but it will also teach the kitten to accept the sort of human handling that most owners want to be able to take for granted with their pets.

BELOW: *Kittens need to be socialized with as many different people as possible, with a range of different ages and appearances.*

One very important factor in maximizing the benefits of handling small kittens is the number of people who interact with them, and this is often one of the most difficult aspects of successfully rearing sociable cats. It has been shown that kittens that are appropriately handled by just one person during their sensitive period of socialization will become very sociable towards that person but they will fail to generalize that acceptance to others. Indeed, research suggests that kittens need to be handled by at least four different people before they will learn to relax

in the presence of people in general, and prospective owners should ask breeders for information about the number of people who have had regular contact with the kittens before making their selection.

Learning about the environment

In addition to learning about cats, people and other species, kittens will need to learn about their environment, and the process of habituation is as important as that of socialization. The hustle and bustle of a family home is probably the best place for kittens to learn about domestic life since here they will be surrounded by the sight and sound of domestic appliances, a diversity of furniture, and a variety of human activities

such as watching television, playing loud musical instruments, decorating and even renovating. This exposure helps the kitten to have a complex view of the world around it and to prepare it for meeting novelty and challenge in its new home. Obviously those kittens that are reared in commercial catteries are unlikely to have been given the same diversity of experience, but this can be overcome by taking litters of kittens between two and seven weeks of age into the breeder's home on a regular basis or by taking a selection of household experiences to them. This may cause a lot of extra work for breeders but it is an essential part of rearing kittens who are destined for life as domestic pets and prospective owners need to look for breeders who take this responsibility very seriously.

Stage 4: Juvenile period

This is the phase of development between the end of the socialization period and the onset of sexual maturity and the precise timing will vary between individuals and between breeds. In a behavioural sense, these kittens are fully developed with a full repertoire of adult responses but during this phase there is a maturation of these responses and kittens will play out their adult roles and learn from experience. Social behaviour starts to consolidate at this time

ABOVE: *Development during the juvenile period prepares kittens for adult life. The length of this period will vary from breed to breed.*

and there is a more obvious division between individuals who seek company and those who prefer a more solitary existence. By the end of the juvenile period the kitten is ready for life as an adult cat and the way in which it responds to its environment and to the people and creatures around it will have been shaped by its early experiences during each of the previous stages of its development.

Understanding and speaking your cat's language

If you were going to share your home on a long-term basis with somebody from another country and culture you would probably take time to find out more about the way in which they viewed the world and make efforts to understand their language. However, very few people who take on another species to be their companion spend time learning about the language of their new housemate or the way in which they interpret the world around them. The cat is just expected to fit into our world and we take communication with our cats for granted. On many occasions, pet and owner muddle through and live relatively fulfilling lives, but for some cats a lack of understanding from their owners can lead to problems and many owners live their lives never getting the most out of their relationship with their cat.

Understanding and speaking your cat's language will help you to see the world through feline eyes, to appreciate the complexity of your pet's communication and to gain a new insight into its behaviour. It will also prevent you from misinterpreting the things that your pet does and will assist in resolving many of the common behaviour problems that cat owners encounter.

BELOW: *Communication between two such diverse species as humans and cats is not always straightforward.*

How cats communicate

Cats communicate with each other and with the world around them in three main ways:

◆ Through vocalization
◆ Through visual signalling
◆ Through the use of smells.

It is the third of these methods that poses the most problems for the cat owner since the human sense of smell is very poor and the scent signals that our cats rely on are lost on us. Visual signals can also be difficult to interpret and when owners have been used to spending time with dogs they will soon find that cats are a very different species. Transferring knowledge of dog communication into the feline world simply does not work and trying to do so can lead to some unnecessary confrontation.

In order to understand how cats communicate it is necessary to understand their social structure. In the wild, cats live together in groups of related individuals and they have very little contact with outsiders. They hunt alone and when they are out on hunting expeditions they aim to limit their interaction with other felines. For these reasons, most of cat language is geared towards keeping your distance and signals that are intended to encourage interaction are usually reserved for members of the same social group.

Keeping it in the family

Cats have been described as the first true feminists and their social groupings are described as being

BELOW: *Wild and feral cats naturally live in related groups, and communication skills are important in maintaining harmony.*

matriarchal. Mothers, daughters, grandmothers, sisters and aunts live together in highly co-operative groups, sharing the rearing of each other's offspring and defending each other from potential intruders. They keep these groups together through the use of certain social behaviours and the most important of these are grooming and rubbing.

Owners of more than one cat will have seen how those that have a good relationship will often cuddle up close to one another and then begin to groom. This grooming is not restricted to the head and neck, which are areas of the body that a cat may find difficult to keep clean, and therefore it is unlikely that this grooming is purely designed to keep

ABOVE: *Related individuals can be extremely relaxed in each other's company.*

the housemate clean! Instead this mutual grooming is used to cement relationships between cats and to exchange important social information in the form of tastes and smells.

As well as grooming one another, most cats will sniff the head and the tail region of their close companions when they meet. It is believed that by doing so they check up on the activities of their housemates by detecting the scents with which they have come into contact while they have been apart.

These scent signals allow the cat to discover what its friend has been eating, where it has been and who

ABOVE: *So, where were you last night? Cats use scent to check where their companions have been and what they have been doing.*

it has been hanging around with!

In addition to grooming, cats also spend time rubbing on one another and this social behaviour is often encouraged by owners in their interactions with their cats. Rubbing is seen between cats who are members of the same social group but its function appears to be very different from the grooming and sniffing which is seen when cats settle down to rest. The most striking thing about rubbing behaviour is that it is usually initiated by the weaker individual. Cats do not live in a structured hierarchy as we do, but they do have respect for one another and while rubbing cannot be classed as a submissive behaviour it does appear to be important as a means of acknowledging status.

This behaviour is also involved in the exchange of scent and when your cat rubs against your legs, he is acknowledging your status, confirming your relationship and also picking up and depositing scent signals. This exchange of scent helps to establish a common signal, which can be used to identify members of the same social group and reassure individuals that they belong.

The use of smell in feline communication

Smell is very important to cats and their bodies are adapted to send and receive scent signals. They have a special organ called the Jacobson's organ, which is used to read scent messages. This organ, which is also found in horses, lions, tigers and other species, is accessed from two small openings in the mouth behind the upper incisor teeth and consists of two fluid-filled, blind-ending sacs which are connected through fine ducts into the nose. A cat that needs to read and understand a scent signal will hold its mouth in a very characteristic position as it sucks the scent up into the Jacobson's organ. This specific behaviour is known as flehmening. Once inside the organ, the scent is concentrated and absorbed and the cat is able to taste the smell in a way that we find very hard to comprehend.

This process enables the cat to gain additional information from the smell and is very important in reading social smells. In addition to possessing special equipment to read scent signals, the cat is also highly adapted to deposit them and there are various parts of the cat's body where special scent-producing glands are found. The scent that the individual produces is unique and helps to identify it to other cats both within its social group and in the wider community. You will have seen your cat using these glands to deposit signals on inanimate objects in the house and garden, on other cats in the household and on yourself.

RIGHT: Rubbing with the side of the face on twigs and leaves is an important way of leaving scent signals for other cats.

The major areas of scent production are the face, the flanks and the tail base, but cats also have glands on the paws, which deposit scent signals during the process of scratching. The depositing of scent signals is called marking, and there are four basic manifestations of this behaviour in the domestic cat:
◆ Rubbing
◆ Scratching
◆ Urine marking
◆ Middening

Rubbing
During rubbing, the cat uses the glands on the face, flank, tail and head to mark items in its environment and individuals which belong in its social group. Cats will often rub their face along twigs in the garden

and you will commonly see this same behaviour in the house when you bring in new items, such as shopping bags and new furniture. You may also have noticed your cat rubbing his chin on your shoes when you sit down; what he is doing is reacting to the multitude of different scent signals that you have brought in on your feet. Rubbing is the most acceptable form of scent signalling as far as owners are concerned and most people will actively encourage their pets to rub around their legs. This full body rubbing is a form of greeting behaviour and is usually accompanied by the erect tail, which signals a desire to interact. It is also a behaviour that is reinforced in most cats by the response of the owners and many individuals use this communication to persuade owners to open another tin of cat food!

Scratching

When cats scratch they are engaging in a very natural and complex behaviour. It is important for cats to keep their claws in trim in order to enable them to hunt successfully, and as the front claws are pulled downwards in that familiar stopping action the blunted outer claw sheath is removed and a glistening new weapon is revealed. Many owners will find these discarded outer claw coverings deposited around their homes and especially at the bottom of favourite scratching posts.

The other functional role of scratching is to condition the muscles and the tendons that are important in moving the claws during the kill and also keeping these important weapons in readiness for use.

As well as the significant role of scratching in caring for the claws, it is also a very important behaviour in terms of communication, and the glands between the pads of the cat's foot deposit a special scent signal as the cat scratches. This scent message is left in addition to a visual signal, which is created by the vertical scratch marks that are left in the surface of the object being scratched.

LEFT: *Cats rub against their owners as part of the greeting ritual. This not only acknowledges your status but also picks up and deposits important scent signals.*

Urine marking

Rubbing and scratching may be thought of as subtle forms of communication but when cats need to get their message across it is common for them to use more obvious signals, including urine. Urine marking is usually performed from a standing position and is called spraying because of the very characteristic position that cats adopt as they back up against the scent post and squirt very small amounts of urine in a horizontal stream onto the vertical surface. They usually have their back slightly arched and will tread with their hind feet while the tip of their tail quivers. Most cats appear to concentrate while they are depositing their signal, and a vacant expression on the face is quite common.

Although spraying is the most common form of urine marking, it is not the only one and some cats will mark with urine that they deposit from a squatting position. Whichever position they adopt, all cats do urine mark at some point in their lives and most do so on a regular basis in their outdoor territory. However, this is not a behaviour that is limited to tom cats, and cats of either sex will still urine mark when neutered.

One of the reasons why scent is so important to cats is that they are solitary predators and therefore when they are hunting they are unlikely to come into direct contact with other cats. Instead they need to communicate with cats who will come along later or who have already passed that way. One of the purposes of this communication is to operate a very elaborate time-share system which will ensure that the available territory is not over-hunted and also minimize the risk of unfamiliar individuals coming into contact and potential confrontation.

Of course, keeping unfamiliar cats apart is just one function of the urine marking and at the other extreme cats that are ready for mating will use the same signals to inform neighbouring cats that they need a mate. In this situation, the urine of in-oestrus females carries an additional message. Tom cats pay a great deal of attention to the marks of these females since they will tell them what stage of oestrus the queen is at and therefore how receptive she will be to the male's advances.

It is this sexual component of spraying behaviour that is affected by neutering, and therefore the likelihood of developing problems associated with this form of marking can be reduced by ensuring that cats of both sexes are neutered.

Middening
As well as using their urine as a marker, cats can use their faeces to communicate with other cats. When they deposit faeces in deliberate locations in order to get a message across to their fellow felines this is called middening. This behaviour is usually seen at the boundaries of the cat's territory and many a neighbour has been upset by piles of cat faeces deliberately deposited in the middle of a well-mown lawn. Some of the locations in which middens are found, such as on the tops of fence posts and on the ridges of roofs, are a testament to the agility of the cat, and many owners find themselves wondering just how their cat managed to deposit its message without falling off!

Vocal communication
Although cats do use vocalization in order to communicate, this is probably the form of feline communication that we know least about. Cats are generally considered to be at their most vocal while they are kittens, and a lot of vocal signalling in the cat is associated with greeting and with social contact. There are thought to be at least sixteen different distinct vocal signals but research is still being carried out into exactly what each of these signals means. The picture is also complicated by the fact that many cats use vocal signals that are unique to them and owners of more than one cat will often comment on the fact that they know which cat is approaching by the type of miaow. It is well recognised that cats are very good at training their

ABOVE: *Vocalization is an extremely effective way for a cat to attract attention.*

owners to respond to their vocal demands and the development of the individual noises is probably connected with the timing of the owner's response. In general terms, cat sounds can be divided into three groups.

◆ **The first group** includes those noises that are produced with the mouth open and gradually closing, in a similar way to our own speech, and examples include the miaow that is used in greeting, and the female and male calling signals used during the mating process. Their aim is to incite social interaction and this group of sounds is associated with amicable encounters.

◆ **In the second category,** there are sounds that are produced with the mouth closed and these include the purr and the trill or chirrup. The purr is a very characteristic feline sound but it is one that has fascinated researchers for years since we are still not absolutely sure how it is produced. It is unusual amongst vocal signals as it continues while the cat is breathing in and out with only a very brief pause being noticed between the breaths. The modern theory relating

to purring suggests that it is produced
by a sudden build up and release of
pressure in the throat which leads to
a rapid separation of the vocal cords.
The rhythm of the vocal cord
movement is generated in the throat
and the frequency of the purr is
always the same. The situations in
which purring occurs are many
and varied and the old myth that all
purring cats are happy is easily
dispelled when you listen to purring
road traffic accident victims and cats
that purr loudly as they are examined

ABOVE: *Totally relaxed with all the fuss, this
Siamese will take all the attention you can give
while purring to show its appreciation.*

on the veterinary consultation table.
This form of communication is
commonly associated with mothers
and kittens and certainly kittens do
use the purr to communicate during
nursing. However, it is also used in
play and during social interactions
with owners, and it appears that the
purr is either associated with periods
of actual interaction between cats or

with people or in situations where social contact is desired. One thing is certain: all cat owners love the sound of the purr and find it very relaxing and rewarding.

◆ **The loud vocal signals** that make up the third category are produced when the cat holds its mouth open in a fixed position, and these have been called strained-intensity calls. Examples include the hiss, the spit, the growl and the snarl and their use is limited to situations of defence and aggression. One specific example of this sort of call is the pain shriek which is designed to startle an attacker into loosing its grip. Anyone who has had to handle a cat against its will understands just why it is classed as a shriek!

Body language

Sound and smell are obviously very important in the feline world of communication but body language is also used to get messages across and owners can gain a lot of understanding by studying how their cat uses both its whole body posture and its facial expressions to communicate. As a result of its solitary predator role, the cat needs to have very clear and unambiguous signals in order to prevent misunderstanding with the strangers it encounters when away from home. The lack of any pack structure means that an injured cat is very vulnerable and therefore most of the cat's communication signals are designed to avoid conflict rather than incite it.

RIGHT: *Ready to defend itself, this Birman makes sure you don't mistake its signals. Its defensive crouching posture is aided by hissing with teeth showing and fur extended but its paw positions show that it is ready to run if need be.*

Body posture

The position of the body and a cat's readiness to fly give us the best indications as to the intention of the cat and although the facial signals are undoubtedly the most important in fine tuning the cat's message it is the body posture that gives the first impression to an approaching cat. Cats are renowned for bluffing their way out of conflict, and raising the hairs along the back and on the tail is often combined with arching of the back and standing sideways on to a potential opponent in order to make the cat look twice its actual size. The theory is that a large cat will scare away any would-be attackers, but in some situations the bluff fails and when this happens the cat will slowly retreat by moving sideways out of range. The slow

ABOVE: *In order to defend itself, this kitten arches its back, fluffs up its tail and tries to look bigger than it really is.*

BELOW: *At close quarters, cats use elaborate visual and vocal signalling to avoid physical confrontation.*

movement is very important in order to prevent inciting the attacker to chase, and the sideways movement allows the cat to keep its adversary in view just in case of trouble!

Bluffing is not always considered an appropriate response and when cats are very frightened they will often shrink to the smallest possible dimensions and try to hide and it is in these situations that the feline maxim of 'I can't see you so you can't see me' really comes into play. All cats would prefer to avoid conflict if possible and this explains why running away is seen as such a desirable option when a cat finds itself in a situation that it does not like.

Facial expressions

When cats are communicating with one another and with us they use their eyes and ears to modify their message, and anyone who has worked with cats will know that the face is the most important part of the cat to watch for clues as to what it is going to do next.

Eyes

Feline eyes are very expressive and the size of the pupil is one indicator as to the emotional state of the cat. Dilated pupils are commonly associated with fear but it is important to read these signals in association with all of the body

ABOVE: *Feline eyes are very expressive but the signals they give need to be interpreted in combination with other body language.*

language that is being displayed since large pupils can also be associated with high arousal unconnected with fear and also with poor light levels! Narrow pupils, on the other hand, are seen as a sign of contentment and, when coupled with a slow blink, the sign is usually of a relaxed and happy cat. Blinking can be used in communication between cats and even from cats to people, and in these situations it is thought to be a way of indicating a level of stress and of seeking reassurance. Staring, however, is the sign of a very assertive individual, and prolonged eye contact is used to intimidate an opponent so it is important to avoid such signals when meeting a cat for the first time. Of course, those people who like cats are more likely to look at them and try to approach while people who do not like cats will usually avoid direct eye contact and narrow their eyes, so it is easy to see how miscommunication can lead to the perverse habit of cats who always approach the people who like them least.

Ears

Ear positions can tell you a lot about your cat's state of mind but people often find them hard to interpret. If they are used to watching dogs there can be some serious confusion since feline ear signals are very different. Ears that are folded sideways and downwards indicate that the cat is trying to avoid confrontation and is preparing to defend itself from an approaching threat, whereas the cat whose ears are flattened against the head with a backwards rotation is getting ready to attack. Ear positions can be altered very quickly and during any encounter it is not unusual to see cats alter their ear position several times as though these small movements are being used to test their opponent's reaction.

The tail end of communication

The tail of the cat is often one of its most striking features, and its role in aiding balance and agility is well recognised. However, the role of the tail in communication is sometimes overlooked and in the past the only comments regarding the tail have related to the belief that the wagging tail of the cat indicates anger and the potential to attack. In fact, the real story is not quite that simple and the

rapid movement of the tail simply indicates that the cat is agitated and is sometimes in a state of emotional conflict. Obviously, this may well lead to the cat attacking if people ignore the signal and continue to interact but this does not necessarily show that the cat is bad tempered. In addition to wagging, cats can use their tails to indicate a range of emotions and to assist in their overall communication.

During greeting, cats will usually approach with their tail in an upright position and the significance of this is usually overlooked. The truth is that a raised tail allows the genital region to be exposed and your cat is inviting you to sniff under his tail and find out all about him! The raised tail signal is usually given before a cat rubs on another cat or indeed on a person and this is important as a means of avoiding conflict. Rubbing is a behaviour associated with relative status and if a cat just waded in and rubbed without asking permission he may find himself in trouble so the raised tail is a friendly gesture used to test out the potential reaction of the other individual and to avoid rejection.

Other tail positions have been associated with sexual communication and, in particular, with signalling of female receptivity, and the bottle brush tail is usually associated with fear and defence. Aggressive cats may also use their tail to indicate their intentions and both the concave and the lowered tail positions are commonly associated with conflict.

BELOW: *A crouched body posture and flattened rotated ears suggest that the cat being approached may lash out if its escape route is blocked. The erect tail of the confident black cat, however, indicates a willingness to interact. Tail posture is an important signal in the feline greeting process.*

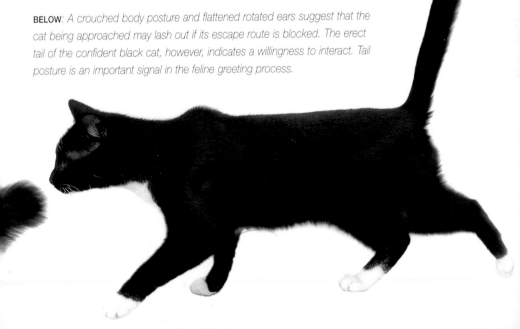

Chapter five

How to be a good owner

In order to maximize your role as a good owner, it is important to look at life from a feline perspective and appreciate how the cat's different perception of the world around it alters its view of the care that we provide. Its enhanced senses of hearing, smell and touch, coupled with its innate desire to hunt, makes it a unique species with very specific needs. As owners, we need to appreciate that the domestic cat is still very close to its wild type in terms of behaviour, and although cats may share their lives with us we simply do not have the same level of control over their existence as we have with our dogs. A consequence of this is that there is an obligation for us to give cats the opportunity to engage in as many of their natural behaviours as possible. In addition, an appreciation of the cat's behavioural, as well as physical, needs is an essential ingredient of good ownership. In all aspects of cat care, studying how cats live in a wild or feral situation can help owners to provide the most suitable conditions for their pet. An understanding of feline behaviour will help owners to make their homes as cat friendly as possible. Cats have traditionally been seen as independent creatures who reluctantly agree to share their lives with humans in return for shelter and food, but in recent years their role as a companion has become more important. This change in emphasis within the cat-owner relationship calls for an appreciation of the differing social needs of our two species. Cats are social creatures but they are also solitary hunters who are ultimately responsible for their survival. As a result, they need to feel in control. Smothering your cat with affection will therefore backfire. The key to a successful relationship is to step back a little, appreciate your cat as a highly adapted solitary hunter and ensure that the daily care you provide enables him to express his natural behaviour.

The world from a feline perspective

One of the problems for owners when trying to see the world from a feline perspective is that the cat's sensory abilities are adapted to a hunting lifestyle in a way that we find difficult to understand.

Hearing
Hearing in cats is very acute – they can detect sounds of up to 60 kHz. This hearing is a direct adaptation for the predatory lifestyle of the cat and it relates to the cat's ability to detect calls emitted by small rodents.

BELOW AND BELOW RIGHT: *The ability to swivel the outer ears, either independently or together, can be useful for the cat when detecting prey.*

Using sound to accurately locate prey enables the cat to stay in hiding right up until the moment of attack and thereby retain an element of surprise. The phenomenal range of movement of the feline ear is another factor in maximizing its hunting ability, and by swivelling the outer ears (pinnae), either together or one at a time, the cat can detect prey through 360 degrees.

Sight
Feline sight is also adapted to the hunter role and although cats are thought to register some colour, their greatest ability in terms of sight is to be able to see in very low levels of light and to be able to detect very

fine movements. The evolutionary explanation is easy to see since cats hunt at dawn and dusk when light is in short supply and their prey are rapidly moving creatures which could be hard to detect. Other adaptations of the cat's eye that are designed to enhance hunting skills include the ability to dilate the pupil widely to maximize the light input and the ability to reflect light from the special layer of cells at the back of the eye, which are called the tapetum. It is this special reflective layer that gives the cat's eye its characteristic night-time glow and which led to the inspiration for making reflective 'cat's eyes' to mark the centres of our roads.

Touch

One sensory ability of the cat which is often over-looked is that of touch. When a cat investigates a new object it will usually do so with its paw before coming closer to touch the object with its nose. The pads are notoriously useless in determining temperature and, as if to illustrate this point, many a cat has walked across the hotplates of an electric cooker. However, they are very sensitive to vibrations and deaf cats are believed to rely on this ability to hear through their feet. Once the initial recce has been made, the cat will then use its nose to find out about the temperature of the object and will use a set of coarse hairs, which are located on the upper lip, around the chin and above the eyes,

ABOVE: *Touch is a very important sense for cats when they are investigating new objects.*

to gather more information. The cat's whiskers are very important in touch and these sensory hairs are characterized by their remarkable mobility. During times of danger the cat will draw the whiskers back for protection but when something is being investigated they are drawn forwards towards it and are used to gain vital information. In the situation where a cat has an item of prey in its mouth, the whiskers are believed to play a very significant role in ensuring that it is in the right position for the cat to perform its fatal nape bite.

Smell and taste

Smell and taste are vital senses for forming a feline view of the world and the importance of scent in communication has been highlighted in Chapter 4. Smell is relatively unimportant in the hunting process and it is the sense of taste that is more likely to be employed during the kill. The feline tongue is known to be very sensitive both to temperature and to taste, but it is remarkably unresponsive to sweet tastes, and this would make sense in view of the somewhat savoury nature of feline prey. Detecting low levels of chlorine and fluorine in water can have a noticeable effect on feline behaviour and leads to the desire of many cats to drink from a puddle in the garden rather than from a bowl of water fresh from the tap. One way of dealing with this is to leave water to settle so that the chemicals have evaporated and resist the temptation to refill the cat's water bowl too regularly. For some cats, any static water is avoided as they seek out sources of running water. This is thought to result from a feline instinct to avoid potentially stagnant water in the wild.

BELOW: *The cat's legendary agility is yet another adaptation to its natural hunting lifestyle.*

Movement

The final aspect of the cat's world that has to be borne in mind when attempting to be a good owner is the need for movement. Cats are true gymnasts but their agile movement is not a game. Good co-ordination, an ability to jump from a standing start and the competence to climb remarkable heights all combine with the skill of being able to walk along perilously narrow ledges to produce a very efficient predator.

The hunting instinct

Once you understand the sensory systems of the cat you cannot fail to realise that the cat is probably one of the best hunting machines ever invented and it is therefore not surprising that most of its behavioural repertoire is linked to its desire to hunt. This is very important in ensuring that you are a good owner.

Hunting for fun?

One very common misconception amongst cat owners is that feeding their cat more will protect the birds and wildlife. In fact, the motivation to hunt has absolutely nothing to do with hunger and this makes very good sense in terms of the cat's need to survive. If a species that hunts alone waits until it is hungry before it tries to detect and dispatch its prey, it runs a very high risk of dying from starvation. After all, prey may simply be unavailable when hunger strikes and there is no pack to feel sorry for the cat and to allow it to share in their kill. The cat is tuned for the kill at all times and when movement and sound combine to trigger the natural instinct, even the best fed and

OPPOSITE: *With its eyes fixed on its prey, this cat is about to use its paw to swat at its unsuspecting victim.*

pampered pet will not be able to resist the desire to pounce.

Playing with prey

Cats use their teeth and their claws in the final dispatching of their prey but the first steps in the hunt are remarkably silent and a hunting cat is a sight to behold. The sequence of eyeing the prey, stalking it and then pouncing is followed by the actions of grasping, biting and killing, and the pattern is not open to variation. Once a potential prey is in sight, the cat is on a mission and will not rest until the deed is done. Of course, there are situations in which the sequence is interrupted and some less competent hunters will never perfect the skills of the grasp, bite and kill despite many repetitions of the eyeing, stalking and pouncing sequence. It is these individuals that are most prone to playing with their prey and to bringing maimed and injured victims back to the security of the home.

Most owners find the sight of their cat playing with its victim very unpleasant and feel that their pet is callous and vindictive, but what they are probably witnessing is the result of an incomplete training in the skills of hunting.

In order to be a good owner, it is very important to appreciate the phenomenal importance of hunting in a feline world and to provide your cat with the opportunity to respond to these calls of nature by hunting and dispatching prey of one sort or another. For many owners, the prospect of their cat contributing to the mass destruction of wildlife is too much and they wish to curtail this behaviour or even bring it completely under control. The options for enabling the owner to live with a clear conscience while still enabling the cat to be true to its heritage are numerous, and most owners will be able to reach some form of compromise.

However, arming the cat with a collar and bell is probably the least effective approach since cats are very intelligent animals and most of them will learn to hold their neck in such a position as to minimize the sound made by the bell.

Reaching a compromise

If you want your cat to have access to outdoors, you can restrict its movement outside at dawn and dusk and during the fledgling season and this can be very effective as has been demonstrated in some parts of Australia where dusk-to-dawn cat

BELOW: *Games that mimic the hunting sequence can provide much needed stimulation for pet cats.*

curfews are the law! Alternatively, you can decide to restrict your cat's movements totally and keep it as an indoor pet, and if this is your decision you need to remember that your cat is now totally dependent on you for the hunting opportunities it needs. Play is therefore vital for these cats and every day they must be offered small rapidly moving targets on which to practice their eyeing, stalking and pouncing.

They should also be given the chance to catch and dispatch some of these prey items and toys must therefore be suitable for this purpose and must not run any risk of fragmenting or breaking when the cat attempts to kill them.

In the wild, cats will spend up to six hours a day hunting Even when you see a cat catching insects, it is not simply playing but is perfecting its hunting skills. Play for indoor cats must always reflect this enormous proportion of the cat's daily time budget and owners of indoor cats need to be prepared to invest some considerable time in this activity.

Indoors or outdoors?

The hunting issue raises the question of keeping cats indoors, but this is not the only reason why more and more owners are opting to keep their pets away from the great outdoors. As with any decision about caring

ABOVE: *Toys that mimic the texture of the cat's natural prey and also provide unpredictable movement can be very attractive to cats.*

for your cat, the over-riding consideration must be the welfare of the cat and thus it is important to consider the issues involved and to make a considered decision about your cat's level of access to outside.

So what are the options?

Although the indoor-outdoor debate is often seen in black and white there are shades of grey, and limiting access to outdoors to certain times of the day is one option, as is building a cat enclosure to allow controlled access to outside or using a harness and a lead to take your cat out into the fresh air on a regular basis.

LEFT: *The car claims many cats' lives each year, and even those cats who appear to have developed some road sense can be at risk.*

One of the main reasons for keeping a cat totally indoors is the risk of injury on the roads. Every year hundreds of cats lose their lives in road accidents and it is not only busy city roads that are the problem. Even relatively quiet country lanes can be a death trap for a young and inexperienced feline, and many owners who have lost their cat in this way are not prepared to take the risk again. Another consideration is disease and many owners opt to keep their pet in an environment where disease risks can be managed.

Disease spread between cats is more likely in a changing population where cats come and go and fights are possible. Infection from poisoned or unfit prey may lead to illness and even death. Encounters with lorries are cited as the greatest danger of outdoor existence, but face-to-face meetings with dogs and foxes can also pose considerable threats. For a truly outdoor cat, there is always the danger of someone else taking him in!

These risks can be minimized by taking the appropriate precautions, such as vaccinating against major infectious diseases, regular worming and restricting the cat's desire to wander and to fight by neutering. Adequate identification will guard against theft, and restricting outdoor access to quiet times of the day can help minimize the risks of fatal encounters with lorries and cars.

Allowing cats outdoors

Although there are valid reasons for indoor life, many people consider access to outdoors to offer the best activity and stimulation for a species that still retains so many of its natural behaviours. However, they should be aware of the risks and how they can help to minimize them.

The indoor alternative

Taking freedom away from a cat that has already had an outdoor life is not a viable proposition since the removal of opportunities to roam and hunt can be very distressing. Kittens who are destined for an indoor life should be kept in from the beginning. Far from being a cruel option, keeping cats totally indoors can be a sensible compromise but it places responsibility on the owner to provide for all of the cat's behavioural needs within the home. An outdoor run is an option which allows the cat the best of both worlds while offering the owner peace of mind.

The indoor cat needs mental as well as physical stimulation; boredom and stress are real dangers if indoor cats are not catered for properly. Lack of exercise can lead to unhealthy weight gain and restricted exposure to new scents can cause a cat to be over-sensitive to changes within the home. As a result of minimal experiences, indoor cats are prone to becoming more dependent on their owners and this may induce over-attachment. Introducing another cat is often more difficult in an indoor household because the resident is not used to encountering strangers' scents and may be highly defensive over its limited territory. Providing adequate stimulation through play and social contact, and offering as many opportunities to engage in normal cat behaviours as possible helps minimize these risks.

Sensible selection is important and the less reactive breeds, such as the Persian, the Birman and the Exotic, are better suited to an indoor existence than the inquisitive and reactive breeds, such as the Burmese and the Maine Coon.

BELOW: *For indoor cats, play is an essential daily activity. Games should supply mental and physical stimulation as well as social interaction.*

Catering for basic feline needs

Although some of the care that cats require from their owners depends on their lifestyle, there are a number of basic feline needs, which every good owner needs to consider and provide for.

Food and water

In order to be a good owner you will obviously need to cater for your cat's nutritional requirements but with the vast array of cat foods that are now available it is impossible to be specific about which type of food to use. The diets that are specially formulated for the differing life stages of the cat can be very useful. It is important to remember that it is not always easy to get the balance of nutrients right

BELOW: *Food is obviously an essential resource but nutritional requirements will alter with age.*

and therefore the preparation of homemade diets can be more difficult than you think.

Cats are obligate carnivores and therefore a vegetarian diet is not a possibility for your pet. Many cats are extremely fussy about what they eat and this sort of behaviour can be avoided by exposing kittens to as wide a variety of foods as possible from an early age.

However, for most owners the kitten stage has passed before the extent of the difficulties are apparent and people find themselves faced with an adult cat that will eat only one or two food types. This can be a problem in view of the cat's unique nutritional requirements, and if your cat is latching onto one food type, such as liver or fish, you will need to seek advice from your veterinary surgeon to ensure that it is getting enough essential nutrients.

If, at any time, you need to alter your cat's diet, then it is important to make the change gradually since any rapid alteration in the animal's diet is quite likely to lead to problems of digestion and clinical signs, such as vomiting and diarrhoea.

Although food is always the topic at the forefront of owners' minds, the provision of suitable drinking opportunities is actually even more important. Cats need access to water at all times and owners need to resist

ABOVE: *Water must be provided at all times, but some cats show a preference for running water and are reluctant to drink from bowls.*

the temptation to change it too frequently (see page 77). Provision of milk is not necessary and although many owners like to offer their cat a saucer of milk a large number of cats are unable to digest it and it can cause quite dramatic gastrointestinal effects!

Who's been sleeping in my bed?

Unlike the dog, cats place very little social significance on their sleeping location and, in the words of the famous poem, 'cats sleep anywhere'. The general ingredients for a favoured resting place are quiet seclusion and warmth and most cats will get out of the way when they want to sleep. An impressive array of cat beds are available on the market and those that offer these important features may be a success but many cat owners spend a fortune on a bed for

ABOVE: *Cats sleep anywhere. A warm, cosy place with familiar smells in a quiet bedroom is often the chosen location for rest rather than an expensive cat bed on show in a busy downstairs family room.*

their cat only to find that it prefers to sleep on top of the boiler. From a behavioural point of view, elevated sleeping areas are likely to be preferred because cats use height to relieve stress and often find being up high very relaxing. Radiator cradles can be good from that point of view as they offer an elevated resting place that is cosy and warm – ideal from a feline perspective. One time when a fixed resting place can be useful

is when a kitten is very young, inquisitive and full of energy, and owners who cannot be present to supervise their new family member at this stage may find the use of an indoor pen advisable. This should not be seen as a prison but rather as a safe haven, and by introducing the kitten very slowly to its use and filling it with interesting toys and activities you can minimize any problems of acceptance.

Healthcare

Caring for your cat's physical health is obviously part and parcel of being a good owner and when you first bring your new kitten home you

should register with a local veterinary practice and take your new pet along for a health check. The day-to-day care of your cat's health will be your responsibility, and regular worming,

vaccinations, teeth-cleaning and flea control will help to keep your pet free from disease.

Neutering is an important part of healthcare for the domestic cat and if your pet is to go outside on a regular basis this is a very effective way of decreasing the risks of injury through fighting and through wandering as

Identification

If your cat is being allowed access to outdoors it is very important to ensure that he has some form of identification. This will allow him to be reunited with you should he become lost and it will also speed up treatment if he is found ill or injured since you, the owner, can be easily identified. The traditional method of cat ID was (and still is) the collar and identification disc and certainly these can be very useful for signalling to would-be adopters that the cat already has a home. Even when a cat lives a totally indoor life, an identification collar is a sensible accessory just in case the door or window is ever left open and the cat manages to escape. When you are considering which sort of collar to purchase you need to look for one that has a safety feature, such as a piece of elastic or a special weakened section of collar which is designed to break if the cat becomes caught on a tree branch or other obstacle. This will ensure that your cat can get free but remember that a broken collar is not much use for ID and therefore the more permanent method of microchipping may need to be considered. The disadvantage of this system is that there is no way of knowing about the ID simply by looking at the cat and for this reason a combination of the two approaches probably offers the cat the best form of protection.

well as helping to keep the cat population under control. It is also a good way of increasing toleration between cats in multi-cat households and of decreasing the risk of anti-social behaviours, such as sexually derived urine spraying. However, neutering is not a panacea for feline behaviour problems and neutered cats can still pose problems for their owners (see Chapter 7).

ABOVE: *Kittens should be introduced to a litter tray at an early age.*

Toilet facilities

If your cat is to be kept as an indoor pet he will obviously need a litter tray, but even outdoor cats need to be kept in confinement when they are very young and will need a litter tray at that time. They may also need to use one at other times throughout their lives, either through illness or circumstances. There are a variety of trays available on the market and it is important to get one that your individual cat accepts. The basic variations between the trays relate to the depth of the pan and to the presence or absence of a hood, and the choice is very much dependent on personal preference.

The type of litter material that

BELOW: *The small, shallow litter tray on the right is suitable for a kitten. Adult cats require a larger, deeper tray (left). Use a plastic scoop to remove any soiled litter.*

is used in the tray can have some very dramatic effects on your cat's behaviour and, in general, the finer litters are better tolerated than the pelleted versions. However, many owners use pellets without any problems and there can be no hard and fast rules. Once the facilities are in the house, kittens may need to be taught how to use them but in most cases the learning process has been completed successfully in the breeder's premises (see Chapter 6).

Grooming

Owners of the long- and semi-long-haired cats will understand the need for regular grooming but short-haired cats will also benefit from being groomed due to the social importance of this sort of owner-cat interaction. Short-haired cats can obviously take care of their own coat in terms of keeping it clean and healthy but cats within social groups also groom one another in order to maintain their relationship and owners can mimic this with a daily session of brushing. Grooming longer-haired cats is also important socially but these cats need help in terms of maintaining a healthy coat, and it is very important that kittens are introduced to the process from a very early age.

Using a brush and comb is not natural for a cat – they tend to use

ABOVE: *Grooming is a very important process, not only for general coat care but also for maintaining social groups.*
BELOW: *If grooming is introduced in a pleasant way at an early age, cats will learn to accept it and only minimal restraint will be required.*

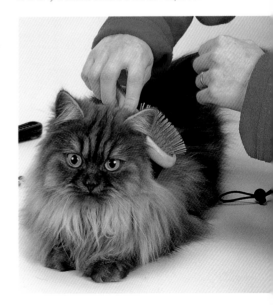

their tongues instead – and therefore failure to introduce these items can lead to fear reactions which can make grooming a very unpleasant experience for both the cat and its owner. Once negative associations have been made it can be very difficult to alter your cat's perception and it is easy to enter a downward spiral in which both the coat of your cat and your relationship with your pet deteriorate rapidly.

The right equipment

The sort of equipment that you use to groom your cat will depend on the coat type and there are a vast array of brushes and combs on the market. A fine metal comb is very useful for removing the loose hair, but do not try to use one to pull out thick tangles since this is likely to be very painful and your cat will soon learn that grooming is no fun. Tangles need to be teased out slowly and gently with a wide-toothed comb whereas the undercoat needs to be tackled with a fine, bristled brush. In the case of short-haired cats and those with unusual coats, such as the Rex breeds, the primary aim of grooming is to remove dirt and loose hairs, and a soft, bristled brush will be ideal for this purpose.

ABOVE: *All long-haired cats gets tangles in their coat from time to time, and these should be teased out gently with your fingers to avoid negative associations with grooming.*

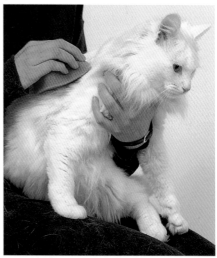

ABOVE: *There is a range of specialized tools that can help in the grooming process. A sponge can be useful for removing dirt from the coats of fine long-haired cats.*

Scratch posts and toys

Play is a very important part of the feline world and kittens need the opportunity to play in order to learn vital adult skills in communication and hunting. The toys you choose for your cat must take into account the natural behaviour of the species and while pet shop aisles are packed with every conceivable toy for cats it is the ones that have been made for cats and not for people that are the best value for money! Often the simple ones are the best, and those that offer unpredictable movement, rapid movement and high-pitched sound are likely to provide your cat with hours of entertainment. The

BELOW: *Playing with your cat is immensely rewarding for you and your pet. Make sure you put aside some time every day to play together.*

ABOVE: *Kittens often make their own toys and this basket has become their unofficial play den. Simple household items often provide them with hours of entertainment.*

BELOW: *Purpose-made toys with a textured feel or those that are impregnated with catnip are popular with many cats. Movement is an important element in feline play and toys that roll can give a cat plenty of quick hunting practice.*

old-fashioned piece of string and rolled-up piece of paper work very well, provided that you are on hand to move them in an unpredictable and exciting fashion, and you can add to the value of this sort of play by rolling the paper downstairs or along ledges and incorporating an element of agility into the game.

Over recent years there has been a trend towards the use of cat aerobic centres and these can be very good value. They incorporate the opportunity for cats to climb, balance and scratch, and many of them have small toys suspended from them, which allow cats to

LEFT: *Play is a social activity and children can have hours of fun playing with their pet.*

Wooden scratching facilities can help to transfer scratching behaviour into an outdoor context and if your cat is destined for an indoor life you need to avoid future confusion by ensuring that the material on the scratch post does not resemble any of your household furnishings.

BELOW: *Feline activity centres combine the features of scratching posts, suspended toys and elevated resting areas to good effect.*

practise their predatory skills. Providing an outlet for these sorts of behaviour is essential for an indoor cat but even when a cat has access to outdoors it can benefit enormously from this type of equipment. Scratching posts provide an opportunity for play but they are also very necessary since scratching is an important behaviour for cats and one that needs to have an acceptable outlet within the home. Probably the most important features of a scratching post are its height, its stability and its surface texture. Tall posts allow cats to scratch at full stretch and the surface material needs to offer a good purchase for the claws.

Catnip and vegetation

A number of cat toys are advertised as being impregnated with catnip. This substance can make the toys very attractive. However, not all cats show a reaction and indeed fifty per cent of the cat population are not responsive to the chemical that is contained in this herb. Those cats that are responsive will experience a period of mild hallucination when the chemical nepetalactone acts on their brain and they will show a short-lived response of excitement which borders on euphoria in some individuals. There is no harm in this response and the chemical is not addictive. Therefore, if your cat is a responder, giving access to catnip can add another dimension to his life.

Grass eating

Many owners notice that their cats will nibble at grass and other vegetation when they are outdoors and these supplements to the daily diet are probably important in aiding digestion and may add to the intake of moisture, roughage and some valuable vitamins. In addition, the power of these substances to induce vomiting can be important in the expulsion of hairballs and therefore cats that are kept permanently indoors should be provided with their own supply of vegetation in the form of windowsill trays of suitable grasses. These are available from most pet shops and will be very useful in preventing your cat from destroying your houseplants!

LEFT: *Chewing on grass is a perfectly normal activity for most cats.*
OPPOSITE: *When a cat is responsive to catnip the effects can be very dramatic (top). When another cat arrives on the scene the catnip can become a much-valued resource, which is worth defending.*

Holiday cover

One important aspect of being a good owner is providing suitable care for your cat in your absence and many owners face the dilemma of whether to leave their cat at home and get someone to look after it or take it to a cattery. There is no easy answer to this debate and, once again, the general rule is to consider what is best for your individual cat.

Cats are usually very strongly bonded to their territory and this can be true for both indoor and outdoor cats but for very different reasons. People often consider a cattery a less desirable option for an outdoor cat which is used to the freedom to explore, but the indoor cat who has only had experience of a very static and limited territory is often more susceptible to the adverse effects of altering environments. Certainly the outdoor cat will be bonded to its home or core territory but will also value its hunting range and the freedom to explore it. However, its access to novelty and challenge will make it more equipped to cope with a temporary change of location than the indoor cat who is bonded to a far smaller area but has limited

BELOW: *Selecting a suitable cattery can be a difficult exercise but the Feline Advisory Bureau will be able to help you.*

capacity to cope with change. On the other hand, the indoor cat is often very strongly bonded to its owner and it can sometimes be better for these cats to go into a cattery than to stay at home where the absence of the owner is very obvious.

Catteries

Catteries offer a very safe and secure environment for cats when their owner cannot care for them and, in addition to their role in caring for cats during holidays, it is worth considering their use at times of any unavoidable upheaval in the household, such as major renovation. This is especially relevant for the indoor cat which might react with behavioural signs of stress when its core territory is severely disturbed.

Selection of a good cattery is obviously essential and you need to look for a place that offers a combination of high-quality care, good hygiene and disease control, and suitable social interaction for your particular cat. The Feline Advisory Bureau publishes guidelines to assist you in your choice but obviously there is no substitute for visiting the premises before you make your final decision.

RIGHT: *It can be reassuring for owners to see their pet settled into its new environment before they leave the cattery.*

Cat sitters

If you do decide to leave your cat at home, it may be worth considering the services of a professional cat sitter rather than relying on a friend or neighbour to pop in and feed your cat. With the best of intentions, friends and neighbours can find it hard to be there at set times and disruption to routine can cause stress in some cats and may even lead to an outdoor cat looking elsewhere for human company and interaction.

Professional sitters, however, will be on hand to care for your cat in much the same way as you do and although they are obviously not owner replacemnets they do offer a very realistic alternative to the cattery.

Chapter six

How and why cats learn

The concept of training is often considered to be irrelevant to cat owners and the independent nature of the cat leads most people to assume that there is little that they can do to bring their pet's behaviour under their control. Training is often associated in the owner's mind with developing certain trick behaviours, and the idea of manipulating a cat to satisfy its owner's selfish whims is understandably unacceptable.

However, there is another far more important aspect to training and that is the development of behaviours that have a definite purpose and will add to the cat's quality of life. Most owners do train their cats on a daily basis, albeit unwittingly in the majority of cases, and common examples of this training include teaching kittens to use a litter tray, introducing the cat

to travel in a cat basket and training cats to come when they are called.

Formal learning in the context of a training class may be something that we associate with our canine companions but this does not mean that cats are immune to learning. Indeed, from the moment they are born, kittens are observing their surroundings, making associations

RIGHT: *Teaching a cat to come to you when it is called can be very useful and is a relatively straightforward process.*

between events and developing behaviours accordingly. As they grow up this willingness to learn can be harnessed and owners who spend time training their cats often find that their relationship with their pet improves as a result. After all, there is no other time when you would engage in such intense interaction and communication with your pet.

Preparing for life

One very important aspect of this learning process is the development of appropriate social behaviour and of adequate coping strategies which will enable the adult cat to cope with novelty and challenge when it meets them. It is important to understand how important the first few weeks of life are in terms of developing these skills and this has been covered in detail in Chapter 3.

Once the foundations have been laid, cats need to learn a variety of behavioural responses and while some of these develop without any need for human intervention there are some skills the cat needs to be taught if it is to get the best out of life in a domestic setting.

BELOW: *Introducing young kittens to a cat carrier can be made easier if you let them observe an adult cat who is already accustomed to the experience of travelling in this way.*

Why are cats so hard to train?

In order to understand the difference between dogs and cats in terms of their receptivity to training, it is important to take into account their different social structure and their widely differing view of motivation.

Dogs are pack animals who need social interaction and co-operation to survive and, as a result, they are keen to comply with the wishes of other members of their pack and see obedience as a way of keeping their pack together.

However, the cat is a social creature that depends ultimately on itself for survival. When it comes to the all-important activity of hunting, the cat is a loner, and at the end of the day the only individual that a cat needs for survival is itself. Of course your cat enjoys your company and even values it but he has no absolute need and if push comes to shove he could cope without you. This differing view of social structure leads to a very different view of co-operation, and when training a cat it is important to realise that your praise and approval are unlikely to be seen as sufficient reward!

Cats expect a decent wage

In order to be successful, training techniques need to be based on reward and this is true for any species, including our own, but what differs is the nature of the reward. A dolphin may work hard for a raw fish and a Border Collie may be spurred on by the prospect of chasing a ball but a cat is unlikely to respond to either of these incentives with any enthusiasm. Determining the best reward for your own cat will involve watching his reactions and finding the things that appear to motivate him most. Some cats will work for games and for social interaction but in other cases the successful feline rewards are food based. Having said

LEFT: *Selecting rewards for cats can be difficult and it is important to ensure that your individual cat sees the treat as something worthwhile.*

LEFT: *Hiding is an important coping strategy for cats, and hostility in the environment will prompt many cats to take cover.*

owner who is hostile and, unlike the dog whose need for social interaction drives it to accept even the most unacceptable forms of attention, the cat will simply move on if its terms are not met. Obviously for the indoor cat moving house is not such an easy option but in these cases cats are likely to respond to punishment by socially withdrawing from the family and leading an increasingly isolated existence.

that, few cats will do anything for their daily cat food and the majority insist on more valuable items, such as cheese, tuna or prawns, before they will comply. There are a variety of cat treats on the market which can be used in training and some cats will work hard to gain access to these, but remember that one important element of reward in cat terms is novelty and many owners have spent a small fortune on cat treats only to find that they cease to work after a relatively short period of time.

Punishment

If you think about the social behaviour of the cat and its relative independence it becomes very obvious that punishment techniques are unlikely to be successful. No self-respecting cat will stay with an

Distraction

In some cases it may be necessary to interrupt a cat's behaviour, and distraction techniques, such as water pistols, have been advocated for this purpose. Certainly distraction can be beneficial when trying to modify a cat's behaviour, but it is important to remember that any link between a hostile experience and the owner is likely to be detrimental to the cat-owner relationship and, since it is impossible to fire water pistols around corners, their usefulness in dealing with cat behaviour is limited. In situations where the relationship between the cat and the trainer is unimportant, for example when a cat is entering someone else's house or terrorizing a neighbour's cat, it may well be appropriate for someone other than the owner to use water as an aversive signal but in the usual cat-owner situation it is rarely appropriate.

Sound can also be useful as a distraction as it can be delivered without any traceable connection with the owner and a sharp hissing sound or a startling clap can be appropriate if a cat is actually caught in the act. However, it is still very important to follow the distraction with a reward for an alternative behaviour, and any cat that has been startled into ceasing a response needs to be shown what to do instead.

Timing

One of the most important aspects of training is the timing of delivery of rewards, and in order to be successful the reward needs to arrive while the cat is actually performing the desired behaviour. Obviously, this can be quite difficult to achieve and anyone who is going to attempt to train their cat will need a lorry load of patience. In general, cat trainers need to be prepared to wait for the desired response to occur so that they can reward it, but this can lead to long time delays and a corresponding drop in owner enthusiasm so there may be occasions when you need to gently encourage your cat to comply. Cats cannot be manipulated into

BELOW: *If a reward is given, it will reinforce the behaviour being performed at that precise moment. Unintentional learning is the cause of many undesirable behaviours, such as jumping on tables and work surfaces.*

responding in quite the same way as the dog and you will need to be very careful about the way in which you interact with your pet in order to keep her attention. Breaking the ultimate goal behaviour down into smaller components can help since you make the individual tasks slightly easier and thereby increase the probability that there will be an opportunity to reward. As each little step becomes established, you can then work to put them all together and eventually achieve your goal.

Litter training

Teaching kittens to use a litter tray is probably the most common example of cat training and yet many owners find that the process is already complete before their new kitten arrives in their home. Indeed the natural cleanliness of the cat is one of the major reasons for its soaring popularity. Cats learn to use litter when they are still with their mother and, although it is now thought that observation plays little if any part in the learning process, it is true that kittens learn to toilet appropriately by having exposure to litter materials from an early age.

When they are born the kittens are totally dependent on their mother and cannot go to the toilet without her assistance but as they become more mobile they start to urinate

and defecate independently. If the mother is good at using a tray, the kittens are likely to follow her and will come into contact with a soft rakeable material which triggers the raking response. Playing in the litter is the first step towards successful litter training, and gradually an association is created between the feel of litter beneath the feet and the act of toileting.

When a kitten arrives in a new home it is helpful for litter facilities to be as similar as possible to those in the previous home and this involves choosing a suitable tray type, litter material and location.

ABOVE: *The scent of the litter tray is an important factor in establishing it as a latrine. Cats should be given adequate opportunity to investigate the smells of any new facilities that are provided.*

The tray should be big enough for the adult cat to comfortably squat, toilet and turn around and it should be deep enough to ensure a good raking action in the litter to cover the deposits.

The decision as to whether to have a covered or an open tray is a very personal one and you may need to experiment with this once your kitten arrives. In general terms, cat litter materials should be fine and easily rakeable but there is a wide variation in litter preference amongst cats and your kitten's personal choice will depend to some extent on the litter that has been used in the previous home. If you do not wish to use the same type of litter it is sensible to make any change over very gradually since the kitten is likely to have developed a preference for the existing substrate.

ABOVE: *Hooded litter trays can offer an extra degree of privacy, especially for nervous cats, but they are not ideal for all cats.*

Positioning the tray

The tray should be positioned in a quiet and secluded location since cats are private creatures who do not like to toilet in areas of busy traffic. They also prefer to keep eating and toileting separate and therefore the tray should be located in an area far away from the feeding bowls.

Problems and accidents

If your kitten does have problems using the tray when he first arrives the best way to help the building of appropriate associations is to put him on the tray when it is most likely that he will need to relieve himself. This will be when he first wakes up, when he has been playing and when he has been fed. Holding his paws very gently you should encourage him to scratch in the litter material, as the action of digging is important to trigger toileting and once he does begin to relieve himself you can gently praise him.

Accidents may happen and it is important to ensure that these are cleaned up effectively so that your kitten does not start to make inappropriate associations with locations in the house and this is covered in more detail in Chapter 7.

For a kitten that is destined for an outdoor life you can train it to transfer its toileting habits from the tray to the garden by mixing litter and soil both within the tray and out in the garden. Tipping the contents of a soiled tray onto a patch of rakeable top soil will help to signal to your kitten that the garden is a suitable toilet and using soil in the tray in the house will encourage the association with an outside substrate.

Cats can do recall too!

Dog owners can appreciate the importance of being able to call their pet to them, but cat owners will also benefit from teaching their pet this simple command. When cats are allowed access to outdoors it is reassuring to know that you have the ability to call them home.

The first step in teaching a reliable recall is ensuring that the cat knows its name and this is done by using it whenever you interact. Cats do not understand English and the only way they can learn the meaning of words is to hear them repeatedly used in the correct context. If you say your cat's name every time you speak to him or look at him, he will quickly come to associate it with reward and with interaction with you. The next step is to make an association between his name and a specific high value reward, such as a prawn or piece of cheese, and once again it is the repetition that is all important. Once the name is established and associated with a

BELOW: *Adopting a crouched posture can help to make you more welcoming and may encourage your cat to come when called.*

reward you can add in a 'come' command, and at this point it is useful to ensure that your cat is already looking at you before you try to interact. This sets you up to succeed since your cat was already thinking of coming your way and you are simply reinforcing his ideas. Slowly but surely you will need to increase the distance between you and your cat and also the level of distraction. You also need to ensure that whenever your cat arrives, however long it takes, you reward him!

Using the cat flap

For cats that are going to be given an indoor-outdoor lifestyle, cat flaps are a very useful invention, although it should be remembered that they are perhaps designed primarily with people in mind. Giving your cat the ability to control his own access to outdoors enables you to go to work in the knowledge that he can get out of the rain and also allows you to continue watching the television while he just pops out through the flap to relieve himself.

For some cats the presence of a cat flap can be disconcerting and this can lead to problems of insecurity (see Chapter 7) but others can adapt very

RIGHT: *As soon as your cat has learnt to use the cat flap, he will have the opportunity to control his own environment.*

well to their presence and use them happily. However, very few cats instinctively know how to react when they first encounter a cat flap and they will need to be trained to use it. As with any training situation you will need to ensure that rewards are used effectively and also that your cat learns that getting through the cat flap will bring access to something worthwhile.

Once the behaviour has been established, the act of entering the house will be rewarded by the comfort of home, and the act of going out will be rewarded by access to outside. However, in the early stages you will need to provide much higher value rewards, such as toys or food. It is important to avoid any situation that might lead to negative associations with the flap and therefore pushing your cat out through the flap

is not a good way to start! In order to begin your cat's association between the cat flap and access in and out of the house it is usually necessary to prop the flap open so that he can actually see the great outdoors and can move through the flap with minimal resistance.

Types of cat flap

There are a variety of different sorts of flap on the market and there are advantages and disadvantages with every model. The solid flaps are perhaps more difficult to introduce than the Perspex variety, because it is difficult for the cat to grasp the connection to outside when it cannot see through the flap. However, some cats find the Perspex flaps challenging because they present a continuous link with outdoors and an opportunity for other cats to look in. The Perspex flaps can be difficult for cats who have limited access to outdoors (for example, being kept in at night) since the flap is never noticeably closed and the cat is likely to try to open it when it is locked and may become frustrated as a result. The solid flap with a locking plate looks different when it is shut and the cat can easily discriminate between when access is available and when it is not.

On the first training session it is usually best to teach the cat to come into the house rather than leave it since the reward of being secure will help to motivate the cat to negotiate the flap. Even so, a food reward or exciting toy may also be needed to reward the cat as he comes through the hole. Once the cat is happily entering and leaving through the open flap, gradually lower it so that he needs to apply some pressure to open it. This part of the procedure can be very slow and you will need to be patient, remembering that loosing your temper or becoming exasperated will only serve to make the process even longer.

Other training opportunities

Although litter training, recall and cat flap use are probably the major examples of cat training there are countless other opportunities for owners to teach their cats to behave in a particular way. Learning can be very useful in preventing behavioural problems and teaching cats to be transported in cat carriers is just one example of this. See Chapter 7 for other examples.

Cat carriers

One of the major problems with cat carriers is that they are only produced when the cat needs to go somewhere. For most cats, the destination on these occasions (usually the veterinary surgery or cattery) is not pleasant. In these circumstances, the cat will quickly develop a negative association with the carrier, and training kittens to enjoy being in their carrying boxes can make these outings far less traumatic.

The first step is to select the right sort of carrier for your cat. Take into account the ease with which the carrier can be cleaned, the ease with which the cat can be put in and taken out of it and the level of security that it offers to your cat. The first two factors are determined by your own perceptions but the third will be dependent on your cat's personality and while some cats seem far more relaxed in a totally enclosed basket others cope much better with one that offers them a view of the outside world.

Whichever basket you decide on, the secret of acceptance is to make positive associations with it and this is best achieved by keeping it on permanent display. Hiding the basket between trips only serves to increase the negative associations. When the basket is on show you can increase the positive image by lining it with a warm blanket and putting cat treats

ABOVE: *Life becomes much less stressful for both cat and owner if your cat learns to go in a cat carrier without any protesting.*

inside for your cat to find. The idea is to let the cat explore the carrier without any interference from you so that he learns that being in it is fun.

When you need to use the carrier to transport your cat to a non-desirable destination this will be seen as a one-off experience rather than the norm. It's also a good idea to spray inside the box with 'Feliway' 30 minutes before you put your cat in as this will help to relax him during his journey.

Problem behaviour

It has been recognised for some time that cat behaviour problems are on the increase, and the number of owners seeking advice about their cat's behaviour has certainly gone up in recent years. In a survey carried out at Southampton University in 2000 the owners of a total of one hundred and sixty one cats were asked about their pet's behaviour and the results were quite unexpected. Half of these cats lived in an urban environment while the other half lived in a rural area but across the board almost all of the owners mentioned some aspect of their cat's behaviour which they were worried about or they simply would rather was not there! Many of these behaviours were perfectly normal from a feline perspective and created a problem because of the context in

which they were being performed. Examples included scratching the furniture, spraying urine on household items and bringing prey back into the house, which was either dead or, perhaps worse, in a semi-dead state. Others appeared to stem from events earlier in the cat's life, which had left the animal prone to the negative effects of stress and unable to deal with novelty and

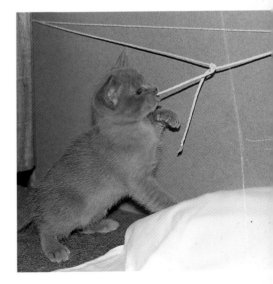

RIGHT: *Although inquisitive behaviour may be considered appealing in kittenhood, the destructive elements of feline behaviour become less acceptable as cats grow older.*

challenge in the world around them.

Many of the situations highlighted in this survey are common everyday problems which are familiar to a large number of cat owners and in a number of cases the problems can be dealt with in a fairly straightforward manner. However, some of the behaviour problems that cats present are not simple and it is always important to remember that there is a strong link between health and behaviour. Any cat that is behaving in an unwanted, unacceptable or unusual manner should be taken to the veterinary surgeon for a proper health check before it starts on a programme of behavioural therapy, and this is especially important in cases of inappropriate house soiling, which is the most commonly reported feline behaviour problem.

Investigating behaviour problems is a complex task and it involves patience, empathy and good scientific knowledge. It is a time consuming business and you can expect to spend two hours or more in a behavioural consultation but it will be time well spent if the motivation for the cat's behaviour can then be accurately identified since this will greatly increase the chances of success in treatment. Obviously this book is not the place for detailed advice on all feline behaviour problems and what follows is a brief look at some of the more common problems.

If you recognise these behaviours in your own cat, or if you have other behavioural concerns about your pet, you should make an appointment to see your veterinary surgeon.

LEFT: *Scratching is a perfectly normal behaviour. However, when it occurs in the house it may be a symptom of a behavioural problem.*

Learning and behaviour problems

The ways in which cats learn was explained in Chapter 6, and an understanding of the feline view of rewards and punishment can be very helpful when you are investigating and treating behaviour problems. Always try to see the situation through your cat's eyes and rely on positive reinforcement techniques for reinforcing appropriate responses rather than punishment for stopping inappropriate ones.

The messy end of life

One of the most common problems cat owners face is the depositing of urine and/or faeces in an inappropriate location in the house. For many households, this behaviour has far-reaching consequences, including tension within families, restriction of the owner's social life and isolation of the cat from the family due to restricting its movement throughout the family home.

The most important thing to establish in these cases is that the cat is in good health and that his behaviour is not related to a medical condition. Once this has been done you will need to determine whether the cat is toileting or using the urine and/or faeces as a marker. This may sound simple but it can take a considerable amount of time to sift through the evidence and reach the right conclusion. Information that will help in this process includes:

◆ The positioning of the cat's urine and/or faeces.
◆ The cat's attitude to the litter facilities you provide.
◆ The posture that the cat adopts when leaving the deposits.
◆ The cat's general demeanour.
◆ Its relationship with other members of the household and the wider community, both human and feline.

One of the most helpful tools in determining the cause of house soiling problems is a plan of the house and garden and an indication of the sites that have been soiled together with the order in which they were used and the frequency with which deposits are found. Often the way in which the behaviour has developed can hold the key to understanding the reasons why.

Indoor marking

In most cases of urine marking the cat will adopt a very characteristic stance as it backs up against vertical surfaces and then sprays urine in a horizontal jet towards them. However, urine marking can also occur with the cat in a squatting posture and

Natural behaviour

Urine marking is an important natural behaviour of cats and is believed to reassure them of their claim on the territory, as well as acting as a timed signal in the elaborate time-share system that they operate. If there is an obvious threat to the cat that can be isolated and dealt with, this can give a very good prognosis for reform of the behaviour, but in all too many cases the urine marking has been going on for some time before help is sought. In these cases, the learned component of the behaviour becomes well established and the feedback mechanism that incites all cats to top up their decaying scent signals comes into action.

cats that are suffering from lower urinary tract disease can adopt a spraying stance in order to pass urine, so the posture of the cat is not enough to determine the motivation for the behaviour. In addition, you will need information about the position of the mistakes, and in marking cases you would expect these sites to be of behavioural significance – for example, around windows and doors and on items that smell strongly of the owner. Cats that mark their territory with urine or with faeces do so in response to

some threat to their territory, whether real or perceived, and in cases where a cat flap is present the perception that there is a permanent breakdown in the home defences can be enough to trigger this very common behaviour. Far from being the work of a confident and belligerent individual these problems are often associated with an insecure cat who is showing other signs of stress-related behaviour and finds it hard to cope with challenge. The deposit is intended as a clear scent signal, and a visual signal in the case of a midden, to those around and also to the perpetrator.

Cleaning up

For the owner the messages contained in urine and faeces are rather lost due to the phenomenally poor human sense of smell and the inability of our species to interpret the social odour signals of our feline companions. This leads to problems in treating these cases and one of the most common reasons for the perpetuation of an indoor marking problem is the inappropriate use of household cleaners to remove the offending scent marks.

Ammonia and chlorine are major constituents of cat urine and when household cleaners that contain these substances are used to clean up, the cat will become increasingly stressed and will be even more prone to engage in unwanted marking.

Instead, mistakes should be cleaned using a regime which will deal with both the protein and the fat components of the urine and remove the scent from a feline perspective. A ten per cent solution of biological washing powder is very effective provided that this is rinsed with cold water and then followed by spraying the area with an alcohol such as surgical spirit. It is important to allow time for the alcohol smell to disperse before allowing the cat access to the area again and it may help to supplement the cleaning with the placing of suitable deterrents in the affected areas. These deterrents should not be confrontational or hostile but rather mimic the feline methods of designating a core or safe territory, which include redefining the area as somewhere to eat, to sleep and to play.

Placing bowls of cat food, pieces of comfortable bedding and a range of suitable toys in the previously soiled areas can be very beneficial, and it can also be useful for owners to deliberately play and interact with their cats when they are in previously soiled areas of the house. The security and familiarity of the house

BELOW: *It's best not to clean the affected area in front of your cat. Disrupting the scent markings can make a cat more anxious, so clean up when the cat's not present. Only allow it access to the area when you have finished.*

can be enhanced by the use of a product called 'Feliway' which is believed to mimic one of the naturally occurring feline smells which is produced by special glands on the face. The odour is believed to reassure the cat of the security of its home and remove the need for the cat to use urine marks. One very important thing to realise is that the cleaning regime already outlined can affect the activity of Feliway and twenty-four hours should elapse between cleaning and applying Feliway to maximize your chances of success. Obviously there is often a lot more to dealing with marking problems than cleaning and redefining the territory and if these simple measures are not effective contact your veterinary surgeon.

Indoor toileting

House soiling which is the result of inappropriate toileting is characterized by cats who refuse to use the litter facilities or become reluctant to go outside in order to relieve themselves. These cats have usually developed some form of aversion to their previous toileting location although some may never have mastered the process of house training. Cats are renowned for their cleanliness and when they begin to leave deposits around the house the strain on the pet-owner relationship can be enormous. This is especially true when the cat has previously been clean and breakdowns in house training account for a large proportion of the feline behaviourist's workload. Finding out why the litter facilities have become unacceptable to the cat is important, but cats do not toilet in areas out of spite or in order to be naughty and therefore it is also important to determine why the cat's new toileting location seems appropriate from a feline perspective.

The substrate

Consider the substrate in each of the locations. Cats prefer soft rakeable materials but owners often find these litter types heavy to carry and opt for the much lighter wooden pellets or newspaper blocks. Unfortunately, the cat has little consideration for its owner's back and if the lighter weight litter is not as comfortable underfoot he will often refuse to use the tray and seek out an alternative. Carpet is an attractive substrate, especially when it is long pile, and before long the cat can establish a toileting association with the floor in the living room! In cases where there has been a recent alteration in the type of litter used, this explanation for the cat's problem behaviour may be the most likely one.

Positioning the tray

The tray's position is also important. When we provide toileting facilities for our cats we should try to mimic the locations they would choose. Trays positioned next to the dog's bed, beside the cat flap, under the stairs and in busy traffic areas in the house are unlikely to be considered acceptable by a species that naturally selects the most quiet and secluded locations to go to the toilet. Equally, trays located close to food bowls pose a problem for the hygienic cat. Since the owner determines where the food is given the cat has no option but to alter its toileting site.

Solving the problem

Simple alterations to litter type and tray location can often be sufficient to make the tray attractive, but it is also necessary to discourage the use of new locations. Do this by ensuring that the areas are cleaned effectively, using the same regime as for marking deposits, and redefining the location as one that is incompatible with toileting. The cat's territory in the wild is divided into core territory and home range with a wider hunting range beyond. Toilet areas are found usually at the boundaries of the core and home areas. For the domestic cat, aim for the house to be interpreted as the core territory, and ensure that activities such as feeding, sleeping and playing are available on a regular and predictable basis. Toilet facilities should be provided at the periphery of that territory, and the cat should then select these rather than soil its core area.

As with marking problems, it is possible for toileting problems to be more complex than is implied here and if the problem has been present for some time it can be frustratingly difficult to resolve. However, the best chance of success comes in cases where an accurate and detailed history has been taken. Behavioural consultations for indoor toileting problems are very worthwhile.

LEFT: *Do not position your cat's litter tray too close to its eating area.*

Scaredy cat

Cats have a confident image and their solitary predator lifestyle makes people think of them as independent and self-sufficient creatures. It is therefore very distressing when a cat develops behavioural problems associated with fear and anxiety and yet these sorts of problems are all too common amongst the domestic feline population. There problems can manifest in a number of ways but the more common examples include:

◆ Over-dependent cats who cannot face life without their owners
◆ Agoraphobic cats who refuse to leave the safety of their home even when total freedom is on offer
◆ Generally fearful cats who run away from visitors, hide from everyday household appliances and even mutilate their own bodies as a result of extreme overgrooming.

The reasons for these problems are complex but in simple terms there is an interaction of genetic input, both from the tom and the queen, and early life experiences which determines the confidence of feline offspring. Those cats that have been deprived of the appropriate levels of socialization and habituation and have the misfortune to be the product of a nervous mother and/or father will be most at risk from these sorts of behavioural problems (see Chapter 3).

Cats will react to stress in one of the two following ways.
◆ Some become very active and obviously distressed and show a desire to escape from the situation.
◆ Other cats become passive and sit as still as they can while attempting

LEFT: *Some cats react to stress by quietly sitting still, keeping out of the way and hoping they won't be noticed, while others will do almost anything to escape the situation.*

to hide from the specific problem.

The first group of cats are likely to be noticed and their owners will often seek advice on how best to help their pet to cope. However, the passive responders are often overlooked, and fear and anxiety in these cats can be a real welfare issue.

Overcoming the problem

Whatever the actual manifestation of fear and anxiety, cats need to be helped to overcome these problems using a combination of desensitization and counter-conditioning techniques as well as positive reinforcement methods to establish new confident responses. Punishment is never appropriate for these cats and over-reassurance should also be avoided. In some cases where the level of fear prevents the cat from learning new behaviours, short-term drug therapy may be needed and your veterinary surgeon will be able to advise you, but such treatment should not be used in isolation and behavioural therapy methods will also be needed.

Grooming problems

Grooming has a number of functions for the cat, including those of coat care, parasite control and social communication but it is its function as a stress-relieving mechanism that can lead to behavioural problems. When a cat finds itself in a state of

ABOVE: *Grooming takes up a large percentage of a cat's day, and keeping the claws in trim is part of the process.*

conflict it will groom itself, and many people have been surprised to see a cat that has narrowly escaped death on the roads sit down on the kerbside and do this. This sort of stress-related grooming is actually perfectly normal. However, when cats find themselves in a state of constant stress or when the stressor that they are reacting to cannot be controlled, the grooming can be taken to extremes. Cats can remove large quantities of hair, often from their flanks, medial thighs and tail head area and, in very extreme cases, they can even begin to self-mutilate by chewing at their skin and muscle tissue.

Treatment of these cases involves increasing the cat's general confidence, limiting or even preventing exposure to the stressor, and providing the opportunities for the cat to use other less damaging coping strategies, such as hiding in high resting places. One very important factor in investigating these cases is ruling out potential dermatological reasons for the hair loss and these cases need a good veterinary work up before they are referred to a behaviour counsellor.

Aggression

In the eyes of the public the subject of aggression is most commonly associated with dogs but when cats begin to show hostile behaviour towards their owners or other cats it needs to be taken seriously. Just because cats are physically smaller than dogs does not limit their potential to inflict injury, and the cat is very well equipped for fighting, with claws at all four corners of its body and teeth at the front! Of course, aggression is a natural response, which is necessary for survival and all animals will show hostile reactions in certain circumstances, but when these behaviours are performed in inappropriate situations they are considered to constitute a behaviour problem.

Biting the hand that feeds you

There are a number of reasons why cats might show aggression to people and, as with any behaviour, it is very important to determine why your cat is behaving in an unacceptable way before you try to treat it.

One of the most common reasons is fear and cats that are showing defensive aggression should be treated with patience and respect. Punishment is never appropriate since it will only

LEFT: *A height advantage can be very important during encounters between cats.*

confirm the cat's fears but you need to remember that the cat has a very strong instinct to run away from danger and therefore finding a reward which is valuable enough to persuade a frightened cat to stick around can be difficult. Lack of appropriate experiences as kittens accounts for a very large percentage of fear-related aggression in cats and it is tragic that such cases are so common when this behaviour is so easily preventable with appropriate rearing.

Hand-reared kittens can present a very characteristic pattern of aggression towards their owners in which they are the perfect pet until something does not quite go their way. During the weaning process, kittens not only learn to cope with a solid diet but also learn to fend for themselves and to cope with situations of frustration. Kittens that have been reared by people are very well socialized and can be very affectionate, but due to a lack of behavioural weaning they have never learned how to cope when things go wrong. Dealing with these individuals involves going back to basics with the weaning process and mimicking the actions of the queen in teaching the kittens to fend for themselves (see Chapter 3).

Cats that suddenly turn on their owners while they are being petted and cuddled can cause great distress, since this contrast between affection and aggression can be quite a shock. Owners will often report that these cats invite interaction by rolling on their backs and exposing their bellies and will accept close physical contact for a while, even purring and showing signs of contentment. Suddenly, however, the mood changes and this sweet little bundle turns into a vicious monster which is sinking its teeth and claws into the owners arm. This behaviour is often referred to as 'Petting and Biting Syndrome', and it is believed that these cats reach a tolerance threshold sooner

ABOVE: *When humans try to make contact with the vulnerable parts of the cat's body, it is perfectly normal for the cat to become defensive.*

ABOVE: *Traditionally, cats are considered to be aggressive to dogs, and most people predict that the dog will come off worse following physical confrontation. However, in most cases cats use their elaborate communication skills to avoid violence.*

than other cats. They genuinely desire interaction with their owners and are often very affectionate between the unexpected bouts of aggression. When we keep cats in our homes we encourage them to behave in a juvenile way and to accept intimate contact which would normally be tolerated only by kittens. However, we also expect our cats to cope in the outside world and interact with other cats as adults. It is thought that Petting and Biting Syndrome is more likely to be displayed by those cats who find it

hard to switch between the roles of independent adult cat and relaxed dependent juvenile. Lack of appropriate handling as a kitten may play a part in this sort of behavioural problem and owners of small kittens are encouraged to pick them up, touch them all over and gently restrain them to prepare them for life as a domestic pet.

Aggression between cats

Most of the cat-to-cat aggression cases that are seen by pet behavioural counsellors involve cats that are living together in the same household. These cases can be extremely distressing for owners; after all, both of the cats involved are much-valued family pets and when problems arise after the introduction of a new pet into an existing household the owner can feel very guilty about disrupting the status quo. In order to understand the hostility that is often shown towards feline newcomers, you need to look at natural feline society and appreciate how this influences your pet's view of the situation. Cats live with other related individuals and their capacity to cope with the arrival of strangers in their midst is very small. The factors that determine how receptive any cat will be to the introduction of another feline include the following:

◆ The availability of resources

◆ The genetic make up of the cat

◆ Its early experiences of interacting with other cats.

If food, warmth and affection are freely available then cats will tolerate the proximity of other cats far more readily and this is well illustrated by the large numbers of cats that congregate around farms and industrial sites.

Introducing a new cat

One of the most important things to consider when introducing a new cat is the importance of scent in the cat's world and the need for the newcomer to integrate its scent into the established family scent profile. This is a slow process but when time is taken to introduce scents, through exchanging toys, bedding and food bowls before the cats meet face to face, they will find it much easier to accept each other.

At the beginning, owners need to separate the two cats and spend time with each of them individually so that they can integrate their reassuring scent with that of the other cat. Giving each cat a piece of the owner's clothing to sleep on can help. When the time comes to introduce the cats, you will need to start the process with some distance between them and slowly bring them together and it can help to do this when both cats are hungry so that they can be distracted with valuable food treats, such as prawns.

Confining one or both of the cats in an indoor pen can prevent the problem of flight-chase cycles where

BELOW: *It is important not to misinterpret interaction between cats – what looks like aggression to us may be part of a playful encounter or it may be simply some much needed maternal discipline.*

fight abscesses and other injuries can turn feline confrontation into full-scale neighbour disputes, and, in some cases, threats and counter-threats can lead to a very unpleasant atmosphere for everyone concerned.

These cases can be difficult to resolve and co-operation between the owners of the victims and the feline bully is essential if progress is to be made. The first step is to set up a time-share system within the neighbourhood whereby the 'problem' cat has access to the outdoor territory at times when the 'victims' are safely tucked up at home. This is an effective way of buying time and enabling the owners to get together in a less distraught frame of mind to discuss what to do next. It is common for despots to pursue their victims into their own homes and attack them there, and treatment in these cases needs to include ways of increasing the attractiveness of the despot's home and making the homes of its victims hostile and unwelcoming.

It is important to use deterrents that won't make the victim's house unappealing to the resident as well. Such owner-driven deterrents as water pistols can lead to a situation where the despot only invades when the

one cat runs and one gives chase, but it will be important to introduce the pens carefully so that the cats see them as a safe haven rather than a prison.

Territorial disputes

Territorial disputes breaking out between neighbouring cats are relatively common, and whilst most of these confrontations will be minor, there are situations where feline despots can inflict a reign of terror over a neighbourhood with disastrous consequences, not only for the cats but also the people. Disputes between owners over who pays the vet's bill for treating recurring cat humans are out of the way. Therefore booby traps or deterrents that cannot be directly traced to the owners are usually more appropriate. Water thrown from an upstairs window can be effective, provided that it is thrown in the vicinity of the cat and not at it. Noise deterrents can also work well if the timing of their delivery is accurate.

Ensuring that the invading cat does not have access to resources such as food could be useful since unintentional reward is often the cause of many repeated incidents of breaking and entering!

Bizarre behaviour

One of the most unusual behaviours that is reported in cats is pica, or wool eating as it is commonly known. This behaviour was first reported in the Siamese in the 1950s but since then it has been identified in Burmese, other oriental breeds and even the moggie! It involves the chewing and eating of materials that are of no nutritional value, and although it is usually referred to as wool eating the targeted materials range from paper, card and rubber to cotton and synthetics. Interestingly, cats that target fabrics either chew or eat their abnormal prey but do not do both, and those that ingest can take in phenomenal volumes of fabric over their lifetime. In some cases, the behaviour can result in blockage of the intestinal tract and the need for abdominal surgery, but others get away with their unusual diet without any medical consequences.

The exact cause of this bizarre behaviour remains a mystery but it is believed that cats are born with a genetic predisposition to behave in this way, which may be related to a miswiring in part of the brain. They need to be exposed to some form of stressor in order for the behaviour to be triggered. The most common age for onset is between two and four months, which often coincides with the time when kittens leave the breeder's premises to live in their new home.

Treatment

This relies on a number of approaches:
◆ Increasing the opportunity for hunting-related play with appropriate toys and food substances
◆ Providing food that requires the cat to chew, such as cooked meat-covered bones
◆ Hiding food around the home to increase the amount of time the cat has to spend searching for its food.

Reducing the attraction of the target material by applying taste deterrents may also help, as can preventing access to inappropriate material, but it is important to give the cat an alternative behaviour which is rewarded. Owner-associated punishment is never appropriate since it will only decrease the cat's willingness to eat these substances in the owner's presence. A secret wool eater is far more difficult to treat!

Summary

The behaviour problems that are encountered by cat owners are many and varied, and while some may represent truly abnormal behaviours, many of them arise out of a lack of understanding between two very diverse species. The social structure of the cat is very different from our own, and feline communication can be difficult to interpret, especially in view of the different sensitivities of the senses in humans and cats.

Any cat that is behaving in an unusual manner should be checked over by a veterinary surgeon, since changes in behaviour can often be an outward symptom of an internal disease state. If the problem is known to be behavioural in origin, owners have the reassurance that it can often be modified or even resolved using the correct application of behavioural therapy. If your cat is displaying behaviour that causes you concern, consult your vet who, if necessary, will refer you to a specialist in the field of cat behaviour. The old adage that prevention is better than cure certainly applies and much can be done to minimize the risks of developing unwanted behaviour. Selecting suitable breeding stock, rearing kittens in a sensitive and suitable manner and taking the time to view life from a feline perspective can all help to foster a better understanding between our two species and, as the cat takes its place as the most popular companion animal, such understanding can only enhance our relationship with this most fascinating of species.

BELOW: *The cat's naturally inquisitive nature can be one of its most endearing qualities.*

Useful information and addresses

There are many organizations that are dedicated to furthering knowledge of cats, and many magazines that are designed to assist cat owners in making the most of their relationships with their cats.

If you need help in selecting a breed of cat, the Governing Council of the Cat Fancy should be able to point you in the right direction. If you require general information about cat care or specific information about feline diseases, the Feline Advisory Bureau will be able to help. It also provides guidelines on catteries and publishes a 10-point guide of what to look for when visiting a cattery. This is designed to help owners select the most suitable holiday home for their pet. The Cats Protection organisation is involved in rehoming but also offers help and support to owners and publishes leaflets and booklets.

If you are the proud owner of a feline companion but you are experiencing difficulties with its behaviour, the first port of call should be your veterinary surgery. Only once your cat has been given a clean bill of health should you seek the advice of a behaviour counsellor. Your veterinary surgeon will be able to refer you to someone who can help. The Association of Pet Behaviour Counsellors offers a nationwide network of behaviour counsellors in the UK, working solely on veterinary referral, as well as members in other countries around the world. A list of UK clinics is available on request.

The Governing Council of the Cat Fancy
4–7 Penel Orlieu
Bridgwater
Somerset
TA6 3PG
Tel: 01278 427575
Fax: 01278 446627
Email: GCCF_CATS@compuserve.com
Website: http://ourworld.compuserve.com/
homepages/GCCF_CATS

The Feline Advisory Bureau
Taeselbury
High Street
Tisbury
Wiltshire
SP3 6LZ
Tel: 01747 871872
Fax: 01747 871873
Email: fab.fab@ukonline.co.uk
Website: www.fabcats.org

The Association of Pet Behaviour Counsellors
PO Box 46
Worcester
WR8 9YS
Tel: 01386 751151
Fax: 01386 751151
Email: apbc@petbcent.demon.co.uk
Website: www.apbc.org.uk

Cats Protection
17 King's Road
Horsham
West Sussex
RH13 5PN
Tel: 01403 221900
Fax: 01403 218414
Email: cpl@cats.org.uk
Website: www.cats.org.uk